Writing a Winning Research Proposal

Writing a Winning Research Proposal

Your Path to Success

M. Rezaul Islam

PETER LANG

Chennai - Berlin - Bruxelles - Lausanne - New York - Oxford

Bibliographic information published by the Deutsche Nationalbibliothek.
The German National Library lists this publication in the German National Bibliography;
detailed bibliographic data is available on the Internet at http://dnb.d-nb.de.

A catalogue record for this book is available from the British Library.

Library of Congress Control Number: 2024031319

Cover design by Peter Lang Group AG

ISBN 978-1-80374-579-4 (print)
ISBN 978-1-80374-580-0 (ePDF)
ISBN 978-1-80374-581-7 (ePub)
DOI 10.3726/b22029

© 2024 Peter Lang Group AG, Lausanne
Published by Peter Lang Pvt Ltd, Chennai, India
info@peterlang.com - www.peterlang.com

M. Rezaul Islam has asserted his right under the Copyright, Designs and Patents Act, 1988, to be identified as Author of this Work.

All rights reserved.
All parts of this publication are protected by copyright.
Any utilisation outside the strict limits of the copyright law, without the permission of the publisher, is forbidden and liable to prosecution.
This applies in particular to reproductions, translations, microfilming, and storage and processing in electronic retrieval systems.

This publication has been peer reviewed.

To Professor Dr Ahmadullah Mia,

In recognition of his unparalleled contributions to the field of social work and social research in Bangladesh.

A visionary leader and mentor, Professor Dr Ahmadullah Mia's unwavering dedication and pioneering research have left an indelible mark on the landscape of social welfare and research in Bangladesh.

With profound gratitude and deep respect, this book is dedicated to his profound influence, guidance, and unwavering commitment to advancing the field of social research in Bangladesh.

His wisdom, mentorship, and unwavering support have been an inspiration for countless scholars and practitioners, paving the way for a more resilient and compassionate society.

Contents

Foreword ... ix

Preface ... xi

Acknowledgments ... xvii

CHAPTER 1
Introduction of a Research Proposal ... 1

CHAPTER 2
How to Make a Suitable Research Title of a Research Proposal ... 19

CHAPTER 3
Writing a Good Introduction to a Research Proposal ... 37

CHAPTER 4
Writing the Problem Statement of a Research Proposal ... 47

CHAPTER 5
Literature Review: Conceptual and Theoretical Framework of a Research Proposal ... 69

CHAPTER 6
Writing Research Objectives and Research Questions of a Research Proposal ... 77

CHAPTER 7
Rationale of the Study of a Research Proposal — 89

CHAPTER 8
Step-by-Step Methodology of a Research Proposal — 111

CHAPTER 9
Research Ethics of a Research Proposal — 135

CHAPTER 10
Scope of the Research of a Research Proposal — 153

CHAPTER 11
Expected Outcomes of the Research of a Research Proposal — 159

CHAPTER 12
Writing Gantt Chart and Budget of a Research Proposal — 169

CHAPTER 13
References of a Research Proposal — 187

CHAPTER 14
Evaluation of a Research Proposal — 193

Bibliography — 213

About the Author — 217

Index — 219

Foreword

In the world of research, a compelling proposal is the cornerstone of success. It is the bridge between your innovative ideas and the resources needed to bring them to life. As someone who has traversed the intricate path of academic and professional research, I understand the profound impact a well-crafted proposal can have on the chance of being funded successfully, which in turn can lead to success in the trajectory of one's career.

Writing a Winning Research Proposal is more than just a guide to writing a successful research grant from conceptualization to evaluation; it is a step-by-step process for success. This book is a beacon for researchers navigating the complex landscape of proposal writing. This book demystifies the research proposal process and makes it less daunting by offering a methodical approach from defining what a research proposal is to the importance of naming and ultimately evaluating the success of the proposal at its conclusion. In other words, the book breaks down the research proposal process into manageable steps. Whether you are a novice just embarking on your research journey or a seasoned professional seeking to refine your skills, this book offers invaluable insights that will elevate your proposal writing to new heights.

The strength of this book lies in its holistic approach. It does not merely teach you the mechanics of writing a proposal but immerses you in the art and science of persuasive communication. The authors, drawing on their extensive experience, provide a wealth of practical examples and exercises that allow you to apply what you have learned immediately. These real-world applications ensure that you are not only absorbing knowledge but also practicing and perfecting your skills.

One of the most compelling features of *Writing a Winning Research Proposal* is its emphasis on impact which connects to the importance of evaluation. In today's competitive research environment, it is not enough to have a good idea; you must present it in a way that commands attention and garners support. This book equips researchers with the tools to do just that, helping you transform your ideas into proposals that stand out and make a difference.

As you turn the pages of this book, you will find yourself guided by the voices of experienced proposal writers who have successfully navigated the challenges you may face. Their best practices and expert advice are like a treasure trove, ready to be mined for your benefit. This book is not just about learning to write a proposal; it is about mastering the craft of proposal writing and using it to achieve your goals.

Embark on your research journey with confidence, armed with the insights and knowledge of these pages. "Writing a Winning Research Proposal" is your key to unlocking the potential of your research ideas and turning them into reality. Whether you aim for academic distinction or professional success, this book will empower you to create compelling proposals that shape the future of your endeavors.

The book will be a well-earmarked tool as you use it time and time again to deliver impactful proposals. This book will assist researchers in delivering impactful proposals equal to the research they seek to support. Enjoy the process!

Patsy Kraeger, PhD
Chatham University

Preface

Welcome to *Writing a Winning Research Proposal: Your Path to Success*. This book is designed to be your comprehensive guide in crafting a compelling research proposal that will pave the way to success in your academic or professional endeavors. Whether you are a graduate student, a researcher, or a professional seeking funding for your research project, this book will provide you with the essential knowledge and practical tips to create a winning research proposal.

Writing a Winning Research Proposal: Your Path to Success is a comprehensive and indispensable guide tailored for researchers, scholars, students, and professionals seeking to master the art of crafting compelling research proposals. In today's competitive academic and professional landscape, a well-crafted research proposal is the key to unlocking new opportunities and advancing one's career. This book stands out due to its holistic approach, providing step-by-step guidance from conceptualization to the final proposal. It goes beyond merely offering surface-level tips and tricks; instead, it immerses readers in the intricacies of proposal writing, equipping them with the knowledge and skills to create research proposals that stand out from the rest.

One of the special characteristics of this book is its emphasis on practicality and application. Throughout the chapters, the author includes numerous real-life examples of research proposals, showcasing different styles, formats, and content. By immersing readers in these practical scenarios, the book enables them to grasp the concepts more effectively and apply them to their unique research projects. To further reinforce learning, interactive exercises, and thought-provoking questions are incorporated within each chapter, encouraging readers to engage with the material actively and develop a solid understanding of the concepts presented.

The book delves deep into critical elements of a research proposal, such as the literature review, methodology, research objectives, and expected outcomes. Each section is meticulously explained, offering comprehensive guidelines for crafting a well-structured and convincing document. Additionally, the author draws upon the expertise of seasoned researchers and proposal writers, infusing the book with valuable insights, tips, and best practices based on real-world experience. This unique blend of theory and practicality ensures that readers not only grasp the theoretical foundations but also acquire the tools needed to create successful research proposals in practice.

Beyond its merits as a resource for proposal writing, this book holds immense usefulness and importance for its readers. For researchers and academics, mastering the art of writing winning research proposals can lead to enhanced research opportunities, increased chances of securing funding, and recognition within their fields. Such recognition can open doors to career advancement, prestigious grants, fellowships, and collaboration opportunities with other distinguished researchers. Moreover, as readers become adept at crafting compelling proposals, they will approach the proposal writing process with newfound confidence and conviction. Armed with the knowledge and skills gained from this book, they can design and articulate impactful research studies that address critical issues and contribute meaningfully to their respective fields.

In the quest for impactful research, academic integrity, and ethical considerations hold utmost importance. The book underscores the significance of proper referencing and responsible research practices, ensuring that readers uphold the highest standards of scholarly conduct. Research proposals built on a foundation of integrity and ethics not only garner respect within the academic community but also generate trust among stakeholders and potential collaborators.

As readers embark on their journey through this comprehensive guide, they are invited to embrace the knowledge and insights it offers. Writing a winning research proposal is a transformative skill that elevates the quality of research and career prospects. The journey to research excellence and success begins with this book as a guiding light. With diligence, dedication, and the wisdom imparted within these pages, may readers create research proposals that leave a lasting impact, advancing their academic or professional pursuits and contributing to the advancement of knowledge in their fields. Happy writing, and may your proposals pave the way for groundbreaking research and enduring success.

The book *Writing a Winning Research Proposal: Your Path to Success* caters to a diverse range of readers, each of whom can benefit in unique ways from its comprehensive content and practical guidance. For graduate students and early career researchers, the book provides a step-by-step approach to crafting compelling research proposals, enhancing their chances of securing funding for their projects and establishing a solid foundation for future research endeavors. Established researchers and academics can refine their proposal writing skills and stay updated on best practices, fine-tuning their proposals for competitive funding opportunities and collaborative projects. The book's emphasis on ethical considerations and academic integrity serves as a reminder to maintain the highest standards in scholarly work.

Professionals and consultants seeking funding for research-based projects or consulting assignments can find immense value in this book. By mastering the art of writing winning research proposals, they can enhance credibility, attract more clients, and secure impactful research and consulting work. The book's cross-disciplinary relevance makes it applicable to various fields, empowering professionals to present their expertise persuasively. Academic and research institutions can use the book as a training resource for their faculty, researchers, and graduate students. Equipping researchers with the knowledge and skills to write effective proposals can increase success rates in securing external funding and research grants. Funding agencies and peer reviewers evaluating research proposals can also benefit from understanding the intricacies of a well-crafted proposal. By aligning with the book's principles, funding agencies can communicate clearer guidelines to applicants, resulting in better-prepared and competitive proposals.

In conclusion, *Writing a Winning Research Proposal* offers valuable insights, practical guidance, and expert advice to elevate proposal-writing capabilities for a diverse readership. From aspiring graduate students to experienced academics, professionals, and funding agencies, the book serves as a transformative resource. By mastering the art of crafting compelling proposals, readers can embark on a path to success, unlocking new research opportunities, and making meaningful contributions to their fields of study and practice. The book's holistic approach, practical examples, and emphasis on academic integrity ensure that readers are well-prepared to navigate the competitive world of research proposal writing and emerge successful in their academic or professional endeavors.

Chapter 1 "Introduction of a Research Proposal" explores the fundamentals of research, research, social research, and research proposal. It covers what a research and research proposal is, its purpose, and its crucial role in the research process. The elements that make a research proposal stand out, including a strong rationale, clear research objectives, and a convincing methodology, will be discussed in detail. By the end of this chapter, readers will have a solid understanding of the importance of a well-crafted research proposal as a roadmap to their research journey.

Chapter 2 "How to Make a Suitable Research Title of a Research Proposal" emphasizes the significance of a captivating research title that encapsulates the study's essence. Crafting an informative, concise, and attention-grabbing title will be the focus of this chapter. Additionally, strategies to align the research title with the objectives and demonstrate the study's significance will be explored.

Chapter 3 "Writing a Good Introduction to a Research Proposal" stresses the importance of setting the stage for the research proposal with a compelling introduction. This chapter will guide readers through the process of writing an engaging introduction, which establishes the research problem, presents the research gap, and outlines the proposal's structure. By the end of this chapter, readers will possess the tools to create an introduction that sparks curiosity and interest.

Chapter 4 "Writing the Problem Statement of a Research Proposal" highlights the critical role of a well-defined problem statement in a strong research proposal. Readers will explore the components of an effective problem statement and learn how to articulate the research problem clearly and concisely. Practical examples and exercises will aid readers in developing a solid problem statement as the foundation for their research investigation.

Chapter 5 "Literature Review: Conceptual and Theoretical Framework of a Research Proposal" emphasizes the significance of the literature review in a research proposal. This section demonstrates familiarity with existing research and theories related to the topic. Readers will learn how to conduct a comprehensive literature review, identify key concepts, analyze theoretical frameworks, and synthesize relevant studies to build a strong conceptual foundation for their research.

Chapter 6 "Writing Research Objectives and Research Questions of a Research Proposal" emphasizes the importance of clear and well-defined research objectives and research questions. This chapter explores the difference between the two and guides the formulation of them effectively. Readers will gain insights into setting specific, measurable, achievable, relevant, and time-bound (SMART) objectives aligned with their research goals.

Chapter 7 "Rationale of the Study of a Research Proposal" explains the significance of the study and its potential contributions to the field. This chapter delves into the rationale-building process, helping readers articulate the research's relevance and practical implications, and persuading stakeholders of the study's value.

Chapter 8 "Step-by-Step Methodology of a Research Proposal" outlines the process of developing a robust methodology for a research proposal. From selecting the appropriate research approach (quantitative, qualitative, or mixed) to designing data collection instruments, readers will gain a comprehensive understanding of the methodological aspects of their research proposal.

Chapter 9 "Research Ethics of a Research Proposal" elucidates the ethical foundation of the study, unraveling the core principles and guidelines that underpin responsible research conduct. This chapter guides researchers through the ethical considerations, emphasizing the importance of informed consent, confidentiality, risk assessment, and the conscientious management of data, ensuring the integrity and credibility of the research endeavor.

Chapter 10 "Scope of the Research of a Research Proposal" emphasizes the importance of defining the scope of the research to set boundaries and focus the study. Readers will learn how to determine the research boundaries, establish the target population, and specify the geographical and temporal aspects of the study.

Chapter 11 "Expected Outcomes of the Research of a Research Proposal" highlights the importance of anticipating the research's expected outcomes to demonstrate its potential impact. This chapter explores how to articulate potential research findings, contributions to knowledge, and applications of the research.

Chapter 12 "Writing Gantt Chart and Budget of a Research Proposal" showcases the value as a project management tool. A Gantt Chart provides an overview of the research timeline and tasks, helping readers plan and schedule their research tasks, allocate resources, and monitor the project's progress effectively. On the other hand, creating the budget for a research proposal involves estimating the financial resources required to carry out the study, encompassing expenses such as equipment, materials, participant compensation, travel, and personnel salaries. The budget provides transparency and accountability, ensuring that the proposed research can be realistically funded and executed within the specified financial constraints.

Chapter 13 "References of a Research Proposal" covers the crucial aspect of referencing. Readers will learn about different citation styles, such as APA, MLA, and Chicago, and how to properly cite sources in both in-text references and end referencing. Proper referencing ensures academic integrity and gives credit to the sources of information used in a proposal.

Chapter 14 "Evaluation of a Research Proposal" offers a comprehensive guide to scrutinizing and assessing the merit of research proposals. This final chapter dissects the key components of evaluation criteria, explores the nuanced process of developing evaluation frameworks, and delves into the peer review process, providing invaluable insights into quantitative metrics and qualitative assessment components for a thorough evaluation of research proposals.

Throughout this book, we have included practical examples, useful tips, and exercises to help you apply the concepts to a specific research proposal. By the end of this journey, you will have the knowledge and confidence to write a winning research proposal that sets you on the path to success in your academic and professional pursuits. Happy writing!

Acknowledgments

I extend my sincere appreciation to the University of Dhaka, Bangladesh, for providing the necessary leave and a conducive working environment that allowed me to dedicate focused time to the completion of this book. This support has been instrumental in the successful realization of this project.

I am also grateful to the Research Institute of Humanities and Social Sciences at the University of Sharjah, UAE, for fostering an intellectually stimulating environment. The resources and collaborative atmosphere have greatly contributed to the depth and breadth of the content within this book.

Special thanks to the two independent blind peer reviewers for their invaluable comments, and to my students and friends for their support, guidance, and insightful contributions throughout the writing process. Your collective efforts have truly enhanced the quality of this work.

To my colleagues and mentors, thank you for sharing your expertise and experiences, and enriching the narrative with diverse perspectives. Your mentorship has been a guiding force.

Finally, to my family and friends, your unwavering support, understanding, and encouragement have been the driving force behind the completion of this book. I am grateful for your presence on this journey.

Warm regards,

M. Rezaul Islam, PhD (Nottingham)
Shaheed Giasuddin Ahmed Residential Area
Dhaka University Campus
Dhaka-1000, Bangladesh
E-mail: rezauldu@gmail.com, rezaul.iswr@du.ac.bd

CHAPTER 1

Introduction of a Research Proposal

ABSTRACT

This introductory chapter lays the groundwork for a thorough exploration of research proposal writing. It elucidates the fundamental concepts of research and research proposals, emphasizing their pivotal roles as precursors to successful research endeavors. Through a nuanced examination, the chapter delineates various types of research proposals, elucidating their unique characteristics and practical applications. Furthermore, the chapter underscores the paramount importance of crafting a well-articulated research proposal, which serves as a roadmap for delineating research objectives, scope, and anticipated outcomes. It systematically dissects the essential components of a comprehensive research proposal, providing a solid foundation for the subsequent chapters. Moreover, the distinction between research proposals and research designs is clarified, highlighting their interconnected nature and distinct attributes. Practical insights are also offered regarding the optimal length of a research proposal, advocating for a concise yet informative approach to convey essential information effectively.

KEYWORDS: Research proposal, types and importance of research proposal, key components of research proposal, research design

1.1 Definition and Features of Research and Social Research

Research is a scientific inquiry, often referred to as a scientific investigation, which examines human behaviors, and their relationships, and delves into social changes over time (Islam, 2022a). Research is an undertaking in which a systematic investigation is conducted to discover the truth concerning a particular question (see Punch, 2016; Creswell & Creswell, 2017; Gastel & Day, 2022; Locke, Silverman & Spirduso, 2019; Leedy & Ormrod, 2019). People worldwide have been endeavoring to explore, analyze, and predict unknown, uncertain, and unexplained phenomena utilizing both theoretical and analytical skills, scientific and non-scientific approaches, as well as indigenous methods and sources of knowledge. As the knowledge of scientific investigation has advanced, individuals have become more rational, logical, systematic, and scientific in their approach to investigating issues to obtain answers (Abusaleh & Anwar, 2022). Several refereed definitions of research are presented as follows:

> "Research is a systematic process of collecting, analyzing, and interpreting-data-to increase our understanding of phenomena about which we are interested or concerned" (Leedy & Ormrod, 2015, p. 20).

> "Research is a process step used to collect and analyze information to increase our understanding of a topic or issue. It consists of three steps: pose a question, collect data to answer questions, and present an answer to the question" (Creswell, 2008, p. 03).

> "Research may be defined as a systematic and objective recording and analysis of controlled empirical observations that may lead to the development of principles, laws, or theories resulting in prediction and possibly ultimate control of events". (Best & Kahn, 1986, p. 04)

Social research is a scientific inquiry focused on social aspects. The primary scope of this investigation encompasses the social perspective, delving into human behavior, cultures, norms, and values, as well as human welfare and services (Khan & Reza, 2022). The following referenced definitions offer various perspectives on the definition of social research:

> Social research is an objective analysis and recording of controlled observations that can help to develop new principles, generalizations, and theories relevant to important incidents in society (Penz, 2006).

> Social research either tests the appropriateness of existing theories that seek to account for the behavior we are interested in, develops new insights, or constructs new theories to help build up our understanding of the process behind this behavior (Henn et al., 2009).

> Social research delves into the constant relationship between social theory and social issues in which both are modified through combinations of reflection, experience, and systematic investigation (May, 2011).

> Social research is a systematic investigation, using the principles of the scientific method to test hypotheses, acquire information, and solve problems on human interrelationships (Barker, 2013).

> Social research is motivated by changes in society and employs scientific ideas to illuminate those changes to demonstrate a scientific interpretation of social change and development (Bryman, 2016).
>
> Social research is a process in which researchers combine a set of principles, outlooks, and ideas with a collection of specific practices, techniques, and strategies to produce new knowledge. Social research is conducted to learn something new about the world, carefully document expectations or beliefs, or refine their understanding of how the social world works. (Neuman & Robson, 2018)

Based on an extensive literature review, Abusaleh and Anwar (2022) have classified research into four main categories: (i) basic research and applied research, (ii) qualitative research and quantitative research, (iii) action research and evaluative research, and (iv) exploratory research and explanatory research. Kumar (2011) categorized social research based on application (such as pure research and applied research), objectives (such as descriptive research, exploratory research, correctional research, and explanatory research), and inquiry (such as qualitative research and quantitative research). I believe that two dominant categories are the objective point of view and the nature of research. From an objective point of view, research can be basic research, applied research, and action research. Basic research is defined as research conducted to test a specific theory or to investigate relations among phenomena, with little or no expectation of applying research results to practical problems. Its main purpose is to increase knowledge in a particular area. "Applied research" is conducted for practical purposes, such as generating findings and recommending long and short-term interventions. It emphasizes solving a specific problem in real situations (Connaway & Power, 2010), and its main purpose is to solve problems. Action research is a research approach that involves the active participation of researchers in a real-world problem or situation. It is characterized by a cyclic process of planning, acting, observing, and reflecting, to address and improve issues in a specific context.

On the other hand, qualitative research is conducted on qualitative phenomena to gain insights into human behavior, motivations, or attitudes (Abusaleh & Anwar, 2022). This form of research is employed in the behavioral sciences to comprehend the motives behind human behavior, such as understanding how individuals behave in specific situations and the reasons behind their actions (Kothari, 2004). In contrast, quantitative research can be characterized as the systematic empirical examination of observable phenomena through numerical, statistical, or computational techniques (Given, 2008). This research approach revolves around the aspect of the quantity or extent of any phenomenon (Mishra, 2017). For instance, it is employed to measure the level of poverty or to comprehend the economic profile of a specific community (Abusaleh & Anwar, 2022).

Social research is of paramount importance as it serves as the compass guiding our understanding of complex social phenomena (Sarantakos, 2013). Through systematic investigation, social research illuminates the intricacies of human behavior, societal structures, and cultural dynamics (Henn et al., 2009). It is the key to unlocking insights that inform policy decisions, drive social change, and contribute to the collective knowledge of humanity. By employing rigorous methodologies, social research not only reveals patterns and trends but also unearths the underlying causes and consequences of social issues and different aspects of social problems to understand their connections and causal relationships (Kumar, 2002). Its significance lies in its capacity to bridge theory and practice, offering evidence-based solutions to contemporary challenges, fostering informed decision-making, and ultimately fostering a more nuanced comprehension of the intricacies of the human experience within the broader societal context.

1.2 The Meaning and Definition of the Research Proposal

In the realm of social research, the journey from understanding the meanings and features of social research leads us to the critical juncture of crafting a research proposal (see Yin, 2009; Myers, Well & Lorch Jr, 2013; Cohen, Manion & Morrison, 2018; Leedy & Ormrod, 2019). This pivotal document serves as the blueprint for our investigative endeavors, guiding us through the intricate process of designing, executing, and interpreting our study. Now that we will delve into the nuances of social research, the next leg of our academic voyage involves translating our understanding into a tangible research proposal. This document is more than a mere formality; it is the roadmap that outlines the trajectory of our research, from inception to fruition.

A research proposal is a meticulously crafted document that reflects the researcher's commitment to exploring and understanding a specific research problem or question. It is more than just a formality; it is the initial steppingstone on the path to scientific discovery and intellectual exploration. As a formal and systematic outline, the research proposal serves as a guiding light for the researcher, charting the course they will traverse to investigate the unknown, find answers, and make meaningful contributions to their field. Creswell and Creswell (2017) emphasize the importance of a well-structured and thoroughly thought-out research proposal as a foundational step in the research process. They highlight that a research proposal serves as a roadmap or blueprint for the study, outlining the research questions, objectives, methodology, and anticipated outcomes. Creswell and Poth (2016) stressed the significance of clarity, coherence, and feasibility in a research proposal, as it not only guides the researcher's own work but also communicates the intended study to others, such as advisors, reviewers, and funding agencies. Additionally, they emphasize the iterative nature of the research proposal, suggesting that it may evolve and be refined as the research progresses and new insights emerge. The process of writing a research proposal is akin to architecting a blueprint for a grand edifice. It requires careful consideration and planning, as every detail counts towards the successful completion of the research project. Just like an architect designs every corner and aspect of a building, a researcher designs every facet of their study in the proposal, ensuring it aligns with their overarching goals.

> A research proposal is a specific kind of document written for a specific purpose. It is because research involves a series of actions and therefore it presents all actions systematically and scientifically. In this way, a research proposal is called a blueprint of the study which simplifies and outlines the steps that the researcher will undertake during the conduct of his/her study. However, a proposal is a tentative plan so the researcher can modify his/her research idea based on his/her reading, discussion, and experiences gathered in the process of research (Islam, 2019).

> A research proposal is a structured, formal document that explains what you plan to research (your research topic), why it's worth researching (your justification), and how you plan to investigate it (your methodology) (Jansen, 2020).

A research proposal is a formal document expressing the details of a research project, which is usually for science or academic purposes, and it's typically four to seven pages long. Research proposals often include a title page, an abstract, an introduction, background information, research questions, a literature review, and a bibliography (Indeed Editorial Team, 2023).

A research proposal is a concise and coherent summary of your proposed research. It sets out the central issues or questions that you intend to address. It outlines the general area of study within which your research falls, referring to the current state of knowledge and any recent debates on the topic. It also demonstrates the originality of your proposed research (University of Birmingham, 2023).

A research proposal is a highly structured document that describes your study's topic and explains how you plan to investigate a specific inquiry. It typically provides an in-depth analysis of the theories that support your hypothesis, which is a projected answer to this inquiry. It can also show which methodologies you plan to use, including the practical steps for conducting your study process. As a research proposal often introduces readers to your topic, this document may also discuss the primary objectives of your project and how it may contribute new information to an academic field. (Indeed Editorial Team, 2023)

In academic, scientific, and professional spheres, the research proposal is a prerequisite for gaining approval and support for the proposed study. Academic supervisors demand it to evaluate the feasibility and relevance of the research undertaken by their students. Funding agencies rely on research proposals to ascertain the merit and potential impact of a research project before allocating valuable resources. Research committees, too, seek well-crafted proposals to assess the viability of research endeavors and allocate institutional resources effectively.

The key to a compelling research proposal lies in its persuasive nature. Every sentence and every word is carefully chosen to persuade the readers that the proposed research is not only significant but also essential for advancing knowledge in the respective field. It is not merely a summary of research intentions; instead, it is an artful composition that instills confidence in the audience about the researcher's capabilities and dedication to achieving their objectives. As an effective communication tool, the proposal transcends mere formality, transforming into a vessel of inspiration. It showcases the researcher's passion for the subject matter, revealing the burning desire to explore uncharted territories and bring new insights to the forefront. It is this passion that compels the reader to invest their time, attention, and resources into supporting the proposed study, believing that the researcher's fervor will lead to significant discoveries and advancements.

A well-structured research proposal is a testament to the researcher's intellectual rigor and analytical prowess. The proposal provides a clear and logical flow of ideas, meticulously presenting the research questions and objectives (Marczyk, DeMatteo & Festinger, 2010). It demonstrates a deep understanding of the existing literature, identifying gaps and opportunities that the proposed research seeks to address. The proposal leaves no room for ambiguity, ensuring that the reader comprehends the scope and intent of the research with absolute clarity. Furthermore, an evidence-based approach lends credibility to the proposal. The researcher supports their claims with empirical evidence, referencing prior studies, data, and established theories that buttress their arguments. By drawing upon credible sources, the researcher proves their competence in the subject area, fostering confidence in the readers that the proposed study is grounded in sound scholarship. A research proposal represents the seed of innovation and discovery. It is the bedrock upon which the researcher's aspirations are transformed into reality. Through its formal and persuasive structure, the research proposal not only charts the course for the research project but also captures the hearts and minds of its audience. As researchers embark on their intellectual odyssey armed with a compelling proposal, they carry the hopes of advancing knowledge and leaving an indelible mark on their field of study.

1.3 Types of Research Proposal

1.3.1 Academic Research Proposal

An academic research proposal is a formal document prepared by researchers, scholars, or students to propose a research project for academic purposes. It is typically submitted to academic institutions, funding agencies, or research committees to seek approval and funding for conducting research. Academic research proposals are common in universities, colleges, and other educational institutions, and they play a crucial role in the research process.

Characteristics of Academic Research Proposal:

- Original Research: Academic research proposals focus on original research projects that aim to contribute new knowledge to the existing body of literature in a specific field or discipline.
- Theoretical Framework: They include a clear and well-defined theoretical framework, outlining the conceptual underpinnings and guiding principles of the proposed research.
- Literature Review: Academic research proposals include a comprehensive literature review to demonstrate the researcher's understanding of the existing research related to the proposed topic.
- Research Objectives and Questions: They present specific research objectives and research questions that the study seeks to address.
- Methodology: Academic research proposals outline the research methodology, including data collection methods, data analysis techniques, and research design.
- Academic Significance: They emphasize the academic significance of the research, highlighting its potential contributions to the field and the broader academic community.
- Ethical Considerations: Academic research proposals address ethical considerations, including informed consent, participant confidentiality, and data protection.
- References and Citations: They follow specific referencing styles (e.g. APA, MLA, and Chicago) to cite sources properly and maintain academic integrity.

1.3.2 Consulting Research Proposal

A consulting research proposal is a document prepared by consulting firms or professionals to present their services and proposed research approaches to potential clients. These proposals are typically submitted in response to a request for proposal (RFP) from organizations seeking consulting services. Consulting research proposals aim to convince clients that the consulting firm has the expertise and capabilities to address their specific research needs.

Characteristics of Consulting Research Proposal:

- Client-Centric Approach: Consulting research proposals are tailored to meet the specific needs and requirements of the client organization.
- Problem Statement: They begin with a clear problem statement or client challenge that the consulting firm aims to address through the research.
- Scope of Work: Consulting research proposals outline the scope of the research project, including the objectives, deliverables, and expected outcomes.
- Proposed Methodology: They describe the proposed research methodology and approach, highlighting how the consulting firm plans to tackle the research problem.
- Timeline and Budget: Consulting research proposals include a detailed timeline for the research project and a budget that outlines the costs of the consulting services.
- Previous Experience and Expertise: They showcase the consulting firm's previous experience, expertise, and relevant case studies to demonstrate their capacity to handle similar research projects successfully.
- Client Benefits: Consulting research proposals focus on the benefits and value that the client will receive from the research findings and recommendations.
- Professional Language: They use professional language and terminology to communicate the consulting firm's expertise and credibility.

While academic research proposals primarily seek funding for scholarly research, consulting research proposals aim to secure a contract with a client to conduct research and provide consulting services. Both types of proposals require careful planning, clear communication, and a persuasive presentation of the research objectives and methodologies.

1.4 Purpose and Importance of Research Proposal

The primary purpose of a research proposal extends far beyond mere formality; it serves as the very cornerstone of a successful research endeavor. At its heart, the proposal acts as a persuasive vessel, aiming to captivate and convince its intended audience of the research's worthiness and significance. Whether the audience comprises esteemed academic advisors, critical funding agencies, rigorous peer reviewers, or invested stakeholders, the proposal must showcase the researcher's capability, enthusiasm, and dedication to the project. It is a golden opportunity for researchers to demonstrate their competence, expertise, and profound understanding of the subject matter. Hansen (2020) mentions that the purpose of the research proposal (its job, so to speak) is to convince your research supervisor, committee, or university that your research is suitable (for the requirements of the degree program) and manageable (given the time and resource constraints you will face).

A research proposal serves two primary functions: firstly, it provides a roadmap for carrying out the research endeavor, and secondly, it aims to persuade either a scholarly institution, a funding organization, or evaluators to grant approval and financial support. While both purposes hold significance, the latter is frequently emphasized in proposal composition, as obtaining permission is a crucial prerequisite for commencing the research (Karim, 2022). By presenting a compelling research proposal, the researcher aspires to foster confidence among the intended audience that the study is not only valuable but also feasible. The proposal is a manifestation of the researcher's commitment, painting a vivid picture of their vision for the research and the potential impact it may have on the broader academic or professional community.

Introduction of a Research Proposal

1.4.1 Define the Research Scope

One of the fundamental purposes of a research proposal is to define and refine the research scope. The proposal serves as a clarion call, ringing out the research questions, objectives, and hypotheses with crystal clarity. It provides a well-lit pathway through the vast landscape of potential investigations, ensuring that the researcher remains on course throughout the journey. By delineating the boundaries of the study, the proposal prevents the research from venturing into uncharted territories or meandering aimlessly, thereby maintaining a strong sense of purpose and focus. In essence, the research scope sets the stage for the entire study, acting as a compass that guides the researcher through the complexities and challenges that lie ahead. By precisely defining the scope, the proposal protects the research from distractions and tangential pursuits, empowering the researcher to stay aligned with the core aims and objectives, and ultimately, to deliver more robust and impactful findings.

1.4.2 Justify the Research

A well-founded research proposal relies on a comprehensive literature review, which serves as a formidable pillar for its justification. By engaging in an exhaustive exploration of existing scholarship, the proposal unearths the gaps and deficiencies in the current knowledge landscape, thus illuminating the need for the proposed research. It is this justification that renders the research proposal a compelling plea to address critical research gaps and enhance the collective understanding of the subject matter. The proposal persuades the readers of the research's value by highlighting the potential contributions it could make to the existing body of literature. The researcher skillfully weaves together evidence from various sources, illuminating how the proposed study has the potential to revolutionize existing theories, shed new light on unanswered questions, or even open up entirely new avenues of investigation.

1.4.3 Secure Funding

In many research pursuits, securing financial support is essential for the realization of ambitious goals. A meticulously crafted research proposal enhances the likelihood of obtaining grants or financial backing from various sources, including academic institutions, government agencies, private foundations, or industry partners. Funding agencies receive numerous proposals, and only the most convincing and compelling ones receive the coveted support. The proposal acts as a powerful advocacy tool, not only showcasing the merits of the research but also providing assurance to potential sponsors that their investment will yield significant returns. By aligning the research goals with the priorities of the funding agency and presenting a clear and feasible plan, the proposal enhances the researcher's chances of securing the much-needed financial resources to embark on their scholarly journey.

1.4.4 Ethical Considerations

An ethically responsible research proposal is a testament to the researcher's commitment to conducting their study with integrity and respect for human rights. Ethical considerations occupy a central place in the proposal, reflecting the researcher's deep sense of responsibility toward the well-being and dignity of research participants. The proposal addresses ethical concerns by outlining how the research will adhere to established ethical guidelines and regulations. It demonstrates that the researcher has thoroughly considered the potential risks and benefits of the study and has taken proactive measures to safeguard the rights and privacy of those involved.

1.4.5 Methodological Approach

At the heart of a research proposal lies the methodological approach, the very framework that underpins the study's design and execution. The proposal meticulously outlines the research methodology, encompassing data collection methods, data analysis techniques, and any potential limitations that may be encountered during the study. By explicating the methodological choices, the proposal not only highlights the rigor and thoughtfulness with which the research will be conducted but also lends credibility to the potential findings. This section showcases the researcher's expertise and understanding of the most suitable methods to address the research questions, thereby instilling confidence in the audience that the study will be conducted with scientific rigor and precision.

1.4.6 Timeline and Resources

In the bustling realm of research, time is of the essence, and resources are valuable commodities. A well-structured research proposal sets forth a realistic and detailed timeline, mapping out the key milestones and phases of the research project. The timeline serves as a crucial organizational tool, ensuring that the research stays on track, and deadlines are met, thereby maximizing the efficiency of the research process. Additionally, the proposal identifies the resources required to undertake the research successfully. This includes not only financial considerations but also access to specialized equipment, materials, databases, or collaborations with experts in the field. By providing a comprehensive outline of the resources needed, the proposal enhances its credibility and demonstrates the researcher's preparedness and strategic planning.

A research proposal is far more than a formal document; it is an intricate tapestry of aspirations, intentions, and meticulous planning. Its purpose is profound—to persuade, guide, and inspire. The proposal serves as a powerful tool for researchers, allowing them to express their passion for their field of study, present a compelling case for the research's significance, and win the trust and support of their audience. By crafting an artful and persuasive proposal, researchers embark on a journey that may not only shape their careers but also contribute significantly to the advancement of knowledge and the betterment of society.

1.5 Key Components of the Research Proposal

A research proposal typically comprises several essential components, including:

- Title: A clear and concise title that reflects the research's focus and objective.
- Introduction: An introduction that provides background information, context, and rationale for the research.
- Literature Review: Conceptual and theoretical framework: A comprehensive review of relevant literature, showcasing the researcher's knowledge of existing research and identifying gaps or unresolved issues.
- Research Questions/Objectives: Clear and specific research questions or objectives that the study aims to address.
- Methodology: A detailed explanation of the research design, data collection methods, sample selection, and data analysis techniques. A discussion of ethical considerations and plans to ensure participant safety and privacy.
- Rationale/Significance and Contributions: An explanation of the research's significance and how it will contribute to the field.
- Scope of study: The scope of study in a research proposal defines the boundaries and extent of the research project, outlining what will be included and excluded. It provides a clear and concise description of the specific research questions, objectives, and target population to be investigated, guiding the researcher in conducting a focused and achievable study.
- Expected Outcomes: The expected outcomes of the study in a research proposal refer to the anticipated results or findings that the researcher hopes to achieve through the investigation. These outcomes serve as measurable and achievable goals, providing a clear direction for the research and helping to assess the success and impact of the study.
- Timeline and Budget: A timeline outlining the key milestones and a budget detailing the anticipated expenses for the project.

A research proposal is a comprehensive document that plays a pivotal role in the research process. It serves as a persuasive tool to convince stakeholders of the importance and viability of the proposed research project. By outlining the research scope, methodology, and significance, a well-crafted research proposal lays the groundwork for successful research outcomes, leading to new knowledge and advancements in various fields of study. Researchers should approach the creation of their research proposals with meticulous attention to detail, ensuring that their ideas are communicated effectively and convincingly to achieve their research goals.

1.6 Differences between Research Proposal and Research Design

Research Proposal and Research Design are two distinct but interrelated components of the research process. While both play crucial roles in shaping a successful research project, they serve different purposes and focus on different aspects of the research endeavor. Here, we explore the key differences between a research proposal and a research design:

Definition and Purpose:
Research Proposal: A research proposal is a formal and systematic document that outlines a researcher's plan to conduct an investigation or study on a particular research problem or question. Its primary purpose is to persuade the intended audience—such as academic advisors, funding agencies, peer reviewers, or stakeholders—that the proposed research is valuable, feasible, and worth undertaking.

Research Design:
Research design, on the other hand, refers to the overall strategy and structure of the research study. It outlines the methods, procedures, and techniques that the researcher will use to address the research questions

and achieve the research objectives. The primary purpose of the research design is to guide the actual execution of the research and ensure that the data collected are valid and reliable.

Focus:
Research Proposal: The research proposal primarily focuses on convincing the audience of the research's significance and viability. It provides a clear rationale for the study, justifies its importance in the context of existing literature, and outlines the potential contributions the research will make to the field.

Research Design: The research design, on the other hand, focuses on the technical aspects of the study. It delves into the specifics of data collection methods, sample selection, data analysis techniques, and other practical considerations required to address the research questions effectively.

Timing:
Research Proposal: The research proposal is prepared before the commencement of the actual research study. It is the initial step, presenting the researcher's plan and seeking approval and support from relevant stakeholders.

Research Design: The research design is developed after the research proposal has been approved. It is the blueprint that guides the researcher throughout the execution of the study.

Content:
Research Proposal: A typical research proposal includes sections such as the introduction (providing background and rationale), literature review, research questions or objectives, methodology overview, ethical considerations, significance and contributions, timeline, and budget.

Research Design: The research design includes a detailed explanation of the research methodology, including the research approach (qualitative, quantitative, mixed methods), data collection methods, sampling strategy, data analysis procedures, and any necessary statistical or analytical techniques.

Level of Detail:
Research Proposal: While the research proposal provides an overview of the research methodology, it does not go into extensive technical details. Its focus is on convincing the audience of the research's merit and feasibility.

Research Design: The research design, in contrast, provides a comprehensive and detailed account of the research methods. It specifies the exact procedures to be followed, ensuring the research is conducted with precision and accuracy.

Flexibility:
Research Proposal: A research proposal is more flexible and open to adjustments. It allows for refinements and modifications in the research design as the study progresses.

Research Design: The research design is relatively inflexible, as it outlines the specific procedures and techniques that should be followed during the research. Deviating from the research design may compromise the validity of the findings.

1.7 How Long a Research Proposal Should Be?

The length of a research proposal can vary depending on the specific requirements set by the funding agency, academic institution, or research supervisor. For example, the University of Birmingham (2023) mentions that a proposal should usually be around 2,500 words. It is important to bear in mind that specific funding bodies might have different word limits. There is no fixed rule for the exact length, but generally, research proposals are relatively concise documents. The typical length of a research proposal is often stated in terms of pages or word count. For many research proposals, including those for academic purposes or grant applications, the length may fall within the following guidelines:

- Undergraduate Level: Research proposals at the undergraduate level may range from five to ten pages, excluding references and appendices. Some institutions may provide specific page or word count limits.
- Master's Level: Research proposals for master's degree programs usually fall within ten to twenty pages, excluding references and appendices. Again, guidelines from the institution or program may dictate the specific length.
- Doctoral Level: Doctoral research proposals are generally more comprehensive and can be twenty to thirty pages or more, excluding references and appendices. Doctoral research often requires a more detailed methodology and literature review sections.
- Grant Proposals: For research grant proposals, the length requirements can vary significantly depending on the funding agency and the complexity of the project. Grant proposals can range from ten to thirty pages or more.

Remember that the quality and content of the research proposal are more important than its length. It is essential to follow any specific guidelines provided by the relevant institution or funding agency regarding the length and structure of the proposal. In any case, it's crucial to be clear, concise, and focused on presenting a compelling research plan and rationale for the study.

A research proposal and research design are distinct yet interconnected components of the research process. The proposal is a persuasive document that seeks approval and support for the research project by outlining its significance and potential contributions. On the other hand, the research design is a detailed blueprint that guides the execution of the research, specifying the methods and procedures to be employed to address the research questions effectively. Together, these components ensure that the research study is well-structured, methodologically sound, and capable of generating valuable insights and knowledge.

CHAPTER 2

How to Make a Suitable Research Title of a Research Proposal

ABSTRACT
This chapter intricately explores the art of crafting effective research titles. It underscores the significance of clarity and precision as fundamental elements for a title's communicative power. Strategies for aligning titles with the research's focal point are meticulously examined to ensure they accurately encapsulate the essence of the study. Additionally, the chapter discusses the importance of incorporating pertinent keywords and phrases to enhance the title's visibility and accessibility. It emphasizes the avoidance of jargon and abbreviations to resonate with a broad audience. The exploration of creativity and impact highlights the role of a well-crafted title in capturing readers' attention without resorting to sensationalism. Moreover, the chapter navigates the crucial phase of reviewing and revising titles to ensure they remain reflective of the evolving research. It suggests seeking inspiration from existing literature and aligning titles with the research's theoretical framework or conceptual model to enhance relevance and coherence. Ultimately, this chapter guides researchers in shaping titles that harmoniously encompass the research's contribution to academia and beyond.

KEYWORDS: Research title, keywords, jargon words, research impact, sensationalism, theoretical framework, research contribution

A well-crafted research title is the gateway to a successful and impactful study. It serves as the face of the research, inviting readers to delve into the depths of the study and explore the researcher's insights. The research title is not merely a combination of words; it is an artful creation that encapsulates the essence of the study and piques the curiosity of the audience. In this chapter, we will explore the essential elements and strategies for creating a suitable research title that captures attention, reflects the research's focus, and communicates its significance effectively.

2.1 Clarity and Precision

The foundation of a suitable research title lies in its clarity and precision, which ensures that the title serves as an accurate representation of the study's core purpose. A clear and precise research title leaves no room for ambiguity, providing readers with an immediate understanding of the research's central theme or subject matter. It acts as a window into the study, offering a glimpse of what lies ahead and enticing readers to delve further into the research. In the fast-paced world of academia and information overload, a well-crafted research title stands out amidst a sea of publications, capturing the attention of potential readers and encouraging them to explore the study further. The title's clarity serves as a guidepost, directing readers to the heart of the research and helping them discern whether the study aligns with their interests or research needs.

To achieve clarity and precision, researchers should avoid using convoluted or complex language in the title. Instead, opt for straightforward wording that succinctly conveys the research's focus without sacrificing its essential details. While brevity is desirable, the title should strike a balance between being concise and providing adequate information. Moreover, a suitable research title should indicate the research's main subject, variables, or phenomena under investigation. It should leave no room for misinterpretation or confusion regarding the research's scope. Additionally, if the research is focused on a specific population, region, or time frame, the title should reflect this specificity, enhancing the title's relevance and appeal to a targeted audience.

Incorporating key terms that are commonly used within the research domain can enhance the title's clarity and assist readers in recognizing its relevance at a glance. This is especially important in interdisciplinary research, where terms from multiple fields may coexist. By incorporating relevant keywords, researchers can align their titles with the vocabulary used by their intended audience, thus increasing the discoverability of the research in academic databases and search engines. Furthermore, a clear and precise research title not only facilitates readers' understanding but also enhances the study's credibility and professionalism. A well-articulated title signals the researcher's attention to detail and commitment to presenting their work in a clear and accessible manner. Such a title reflects the researcher's expertise and conveys the impression of a well-structured and thoughtful study.

Clarity and precision are fundamental attributes of a suitable research title. A well-crafted title acts as a beacon, guiding readers to the heart of the research and providing a glimpse of the study's essence. By avoiding ambiguity, using straightforward language, and incorporating relevant keywords, researchers can ensure that their research title communicates effectively and invites readers to embark on a rewarding intellectual journey. A clear and precise research title is a powerful tool that not only captures the attention of potential readers but also sets the stage for a compelling and impactful research study.

2.2 Reflecting on the Research's Focus

An essential attribute of a suitable research title is its ability to accurately reflect the research's focus and core objectives. The title acts as a concise summary of the study, encapsulating its central theme, research questions, or key variables. It serves as a powerful tool to convey the essence of the research, enticing potential readers to explore the study further and understand its significance. To achieve this, the research title should align closely with the research's intended contribution to the field. It should communicate the primary aim of the study, whether it is to investigate a specific phenomenon, explore a novel theory, test a hypothesis, or analyze empirical data to draw meaningful conclusions. The research title is not merely a formality; it is a reflection of the research's purpose and the researcher's scholarly intentions.

One effective approach to crafting a research title that truly reflects the research's focus is to identify the main research question or the central hypothesis. The title can then be framed around this key inquiry, drawing attention to the heart of the study. For instance, a research study seeking to explore the impact of climate change on biodiversity might have a title like "Assessing the Biodiversity Decline: Implications of Climate Change on Ecosystems." Furthermore, the research title should consider the level of specificity appropriate for the study. Depending on the research's scope and complexity, the title can either be broad and general or more specific and targeted. A study with a broad focus may employ a title that captures the overarching theme, inviting a wider audience to engage with the research. On the other hand, a study with a narrow scope or specific research population may benefit from a more focused title, appealing to a particular set of readers with a direct interest in the topic. In addition to reflecting the research questions or objectives, the research title can also convey the theoretical framework or conceptual model guiding the study. This can provide readers with valuable context and offer a glimpse into the theoretical underpinnings that inform the research's approach and methodology.

Importantly, the research title should be crafted with the target audience in mind. Researchers should consider the interests and preferences of their intended readership when framing the title. A title that resonates with the intended audience is more likely to attract attention and generate interest in the study. Ultimately, a research title that successfully reflects the research's focus not only helps readers grasp the study's central theme but also demonstrates the researcher's clarity of purpose and scholarly acumen. It serves as a signpost, directing readers to the core of the research and inviting them to explore the study's findings and implications.

Crafting a research title that accurately reflects the research's focus is crucial for conveying the study's essence and attracting the attention of potential readers. By aligning the title with the research questions or objectives and considering the level of specificity and the target audience, researchers can ensure that their research title serves as an effective gateway to the study's insights and contributions. A research title that truly reflects the research's focus is a key element of a successful and impactful research endeavor, setting the stage for a rewarding and enlightening intellectual exploration.

2.3 Keywords and Phrases

Integrating relevant keywords and phrases in a research title is a pivotal aspect of modern academic publishing. In the digital age, where research databases, online journals, and search engines have become primary tools for accessing scholarly content, the strategic use of keywords enhances the visibility and discoverability of research studies. A suitable research title should act as a beacon, guiding potential readers to the study by accurately representing its core themes and subject matter through the incorporation of essential keywords.

The process of selecting appropriate keywords begins with careful consideration of the research's main concepts, variables, and themes. Researchers should identify the key terms that best encapsulate the study's focus and are widely used within their field of research. These terms should reflect the core content and context of the research, ensuring that the title aligns with the vocabulary commonly employed by scholars in the domain.

By strategically integrating relevant keywords, the research title becomes more likely to appear in search results when readers explore related topics. A well-optimized title has a higher chance of catching the eye of researchers, practitioners, and students seeking literature on the subject. Consequently, this increases the study's potential for attracting a broader readership and generating more impact within the academic community. However, while keywords are essential for visibility, researchers must strike a delicate balance. The goal is not to overload the title with an excessive number of keywords or to compromise the title's clarity for the sake of optimization. The research title should remain concise and informative, effectively conveying the essence of the study while incorporating relevant keywords in a natural and contextually appropriate manner.

It is crucial to keep track of trends and emerging terminology within the research field. Language and terminologies evolve, and new phrases or concepts may gain prominence. Staying up-to-date with these developments allows researchers to adapt their keywords to better resonate with the current state of the discipline. Additionally, researchers should be mindful of the preferences and conventions within their specific research community.

Different disciplines or subfields may have unique terminologies or variations in keyword usage. Understanding the language preferences of the target audience helps in optimizing the title to effectively reach the intended readership. The process of selecting keywords and phrases should be an ongoing endeavor. As research progresses, new insights and findings may emerge that warrant updates or refinements to the research title. Regularly revisiting the title and assessing its alignment with the research content and audience expectations ensures that the study's visibility and relevance remain optimal.

The strategic use of keywords and phrases is a vital aspect of crafting a suitable research title in the digital era. By carefully selecting relevant terms that accurately represent the study's central themes and incorporating them organically into the title, researchers can significantly enhance the research's visibility and discoverability. A well-optimized research title serves as a gateway, attracting readers and encouraging them to explore the study further, thus maximizing the research's potential impact and contribution to the scholarly community.

2.4 Avoiding Jargon and Abbreviations

An important aspect of creating a suitable research title is the avoidance of jargon and excessive use of abbreviations. While academic research often involves specialized terminology and acronyms within the body of the study, the research title should be accessible to a broader audience. The title serves as the first point of contact between the study and potential readers, and its primary purpose is to convey the research's focus clearly and concisely. Jargon refers to technical language or terminology that is specific to a particular field or profession. While it may be appropriate within the context of the research itself, using jargon in the title can create barriers to understanding for readers who are not familiar with the specific discipline. A research title laden with obscure terms may alienate potential readers from other fields, making them less likely to engage with the study. Researchers should prioritize clarity and universality in the title to ensure that it remains accessible and understandable across diverse academic audiences.

Similarly, an overreliance on abbreviations in the title can hinder comprehension. While abbreviations may be convenient within the research community, they may not be immediately recognizable or intuitive to readers from other disciplines. If abbreviations are essential to the research title, researchers should ensure that they are well-known and widely accepted within the academic realm. Otherwise, it is best to avoid them and use full terms instead. The avoidance of jargon and abbreviations in the title is not a compromise on academic rigor or precision. On the contrary, it is a demonstration of effective communication and consideration for the broader readership. The title should provide a clear indication of the research's central focus, using language that is accessible to both subject matter experts and non-experts alike.

Choosing language that is clear and universally understood does not imply diluting the research's complexity or depth. A well-crafted title can succinctly capture the essence of the study without sacrificing its academic rigor. Researchers should aim to strike a balance between simplicity and accuracy, ensuring that the title remains informative while maintaining its appeal to a wide audience. To achieve this balance, researchers can enlist the help of colleagues or peers from different disciplines to review the title. Feedback from individuals outside the immediate research field can provide valuable insights into how well the title communicates to a broader audience. Moreover, researchers can use the title to create intrigue and interest without resorting to jargon or abbreviations. Crafting a captivating and thought-provoking title can stimulate curiosity and encourage readers to explore the research further, irrespective of their academic background.

Avoiding jargon and excessive abbreviations in a research title is a crucial element in ensuring the title's accessibility and appeal to a diverse readership. By prioritizing clarity and universality in the language used, researchers can effectively communicate the research's focus and significance to a broader audience. A well-crafted research title strikes the right balance between simplicity and academic rigor, inviting readers from various disciplines to engage with the study and discover its valuable insights and contributions.

2.5 Creativity and Impact

In the realm of academic research, where scholarly publications abound, a research title that stands out from the crowd is a powerful asset. Creativity in crafting the research title can make a significant difference in capturing readers' attention and generating interest in the study. A creative and engaging title serves as an invitation, beckoning readers to explore the research further, and can be a distinguishing factor that sets the study apart from others in the field. While precision and clarity are paramount, researchers should not shy away from injecting creativity into their research titles. A creative title has the potential to evoke curiosity, intrigue, and even emotional resonance with the readers, thereby kindling their interest in the study's findings. The research title serves as the face of the research, and a creative approach can add a layer of appeal and allure to draw readers into the study's narrative.

The creative elements in the research title may take various forms, such as using literary devices like metaphors, alliteration, or wordplay, or presenting the research's central concept in an unexpected or thought-provoking way. However, it is essential to ensure that the creativity employed in the title aligns with the research's subject matter and does not compromise the title's accuracy or integrity.

Balancing creativity with precision is a skill that researchers can cultivate through thoughtful iterations and feedback. Seeking input from colleagues, mentors, or peers can provide valuable insights into the impact and appeal of the title. Additionally, conducting a thorough review of titles from other studies, particularly those considered groundbreaking or influential, can serve as inspiration for creative approaches that have proven successful in the academic landscape.

Moreover, the creativity of a research title is not solely about the use of literary techniques; it can also stem from the innovative nature of the research itself. If the study is exploring a novel theory, proposing groundbreaking findings, or challenging conventional wisdom, the title can reflect this innovation by highlighting the study's unique contribution. A title that alludes to the research's pioneering nature can intrigue readers and attract attention from researchers interested in cutting-edge advancements in the field.

While creativity is encouraged, researchers should exercise caution in avoiding overly sensational or misleading titles. The research title should remain truthful and authentic, accurately representing the study's objectives and findings. Sensationalism can lead to disappointment among readers if the research fails to live up to the grand claims made in the title. Instead, creativity should enhance the research's appeal while maintaining academic integrity.

Creativity plays a crucial role in crafting a research title that captures attention, generates interest, and sets the study apart in a crowded academic landscape. By skillfully incorporating creative elements, researchers can create titles that resonate with readers, evoke curiosity, and invite exploration of the study's insights. The research title should strike the perfect balance between creativity and accuracy, leaving a lasting impression on readers and elevating the study's potential for impact and recognition within the scholarly community. A creatively crafted research title is a testament to the researcher's dedication to effective communication and scholarly innovation, paving the way for an intellectually enriching journey through the study's findings and contributions.

2.6 Avoiding Sensationalism

While creativity can be a valuable asset in crafting a research title, researchers must be cautious to avoid falling into the trap of sensationalism. Sensational titles are those that use exaggerated language or grandiose claims to attract attention, often at the expense of accuracy and integrity. While sensational titles might initially draw readers in, they can lead to disappointment and disillusionment if the research does not live up to the promises made in the title. The primary purpose of a research title is to provide an honest and accurate representation of the study's content and objectives. It should clearly and concisely convey the research's focus and main findings without resorting to hyperbole or exaggeration.

Sensationalism undermines the credibility of the research and erodes trust among readers and the academic community. In an increasingly competitive academic landscape, researchers might feel pressure to make their titles more attention-grabbing to stand out. However, it is essential to remember that the true impact of a study lies in the quality of its research design, methodology, analysis, and findings, rather than the flamboyance of its title.

Researchers should resist the temptation to overstate the significance of their findings in the title. Instead, they should adopt a balanced and measured approach that accurately reflects the study's contributions and potential implications. A title that makes modest yet meaningful claims is more likely to garner respect and credibility among peers. One way to avoid sensationalism is to rely on evidence-based language in the title. If the research's findings are supported by robust empirical evidence, the title can reflect this by emphasizing the study's solid foundation in data and analysis. Moreover, researchers can seek feedback from colleagues or mentors during the title crafting process to gauge whether the title strikes the right balance between impact and accuracy. Constructive criticism can help identify any elements of sensationalism or overstatement and guide researchers toward crafting a more authentic and reliable title. While it is essential to generate interest in the research, researchers should do so without compromising the study's scientific integrity. The research title should be an honest reflection of the research journey and its discoveries, leading readers to explore the study with genuine curiosity.

Avoiding sensationalism in the research title is paramount for maintaining the research's credibility and integrity. A research title should be a truthful and authentic representation of the study's objectives and findings, avoiding exaggerated language or grandiose claims. Researchers should prioritize accuracy, transparency, and evidence-based language to ensure that the title reflects the research's genuine contributions to the field. A well-crafted research title, free from sensationalism, not only fosters trust and respect among peers but also sets the stage for readers to engage with the study's valuable insights and meaningful impact.

2.7 Reviewing and Revising

Crafting a suitable research title is not a one-time task but a dynamic and iterative process that benefits from continuous reviewing and revising. A well-crafted title can significantly impact the research's reception and visibility within the academic community. Therefore, researchers should invest time and effort in honing the title to ensure it accurately represents the research's focus, appeals to the intended audience, and aligns with the study's contributions.

Upon initially formulating the research title, researchers should critically evaluate whether it effectively conveys the study's core theme and objectives. Does the title provide a clear and concise indication of what the research is about? Does it capture the essence of the study's findings or theoretical framework? These questions should guide the evaluation process to ensure that the title aligns seamlessly with the research's content and purpose.

Feedback from peers, mentors, or colleagues can be invaluable during the title-crafting process. Seeking the perspectives of individuals outside the immediate research team can provide fresh insights and alternative viewpoints. These external perspectives can help identify any potential issues with clarity, focus, or creativity in the title, prompting researchers to make necessary adjustments to enhance the title's impact. Additionally, researchers can conduct informal surveys or seek opinions from individuals who are not experts in the field to gauge how accessible and engaging the title appears to a broader audience. This feedback can reveal whether the title is effective in generating interest among non-specialist readers and, if necessary, guide researchers in refining the wording to strike the right balance between appeal and accuracy.

As the research progresses and findings unfold, researchers might discover new dimensions or implications that were not initially evident during the title formulation. This discovery phase offers an excellent opportunity for revision. A revised title that incorporates emerging insights ensures that the title remains relevant and up-to-date with the study's current trajectory. Furthermore, researchers should keep an eye on trends within their research domain. The field of knowledge is dynamic and continuously evolving, and new terminology or concepts might emerge over time. Being attuned to the evolving language and research trends can help researchers optimize their titles for relevancy and resonance within the contemporary academic landscape. Maintaining a record of potential title options and variations can be useful during the revision process. Experimenting with different wordings and approaches allows researchers to explore diverse angles and find the most compelling iteration of the title. Keeping track of these options also ensures that researchers have a comprehensive view of the title's evolution and the decision-making process.

Reviewing and revising the research title is a crucial aspect of the title-crafting journey. An iterative approach allows researchers to refine the title, ensuring it accurately reflects the research's focus and potential contributions while resonating with the intended audience. Seeking feedback from peers and non-specialists, staying abreast of evolving trends, and documenting potential variations all contribute to creating a research title that is both impactful and enduring. A well-revised and thoughtfully crafted research title serves as a compelling introduction to the study, inviting readers to explore the research's insights and discoveries with heightened curiosity and anticipation.

2.8 Seeking Inspiration from Existing Literature

Drawing inspiration from existing literature can be a valuable and enlightening approach when crafting a suitable research title. Research titles from reputable sources can serve as valuable examples, providing insights into effective wording, structure, and strategies that have proven successful in the academic landscape. By reviewing titles of related studies and influential publications, researchers can gain inspiration and guidance in formulating a research title that resonates with their intended audience and aligns with the scholarly conventions of their field. One of the primary benefits of seeking inspiration from the existing literature is that it offers a benchmark for researchers to gauge the standards and expectations in title construction within their field. Different disciplines may have varying traditions and norms concerning the style, language, and emphasis of research titles. By analyzing titles from leading journals and influential works, researchers can gain a deeper understanding of how to craft a title that complements the disciplinary context while still retaining its uniqueness and originality.

Existing research titles can also provide valuable insights into the diversity of approaches and creative possibilities in title formulation. By exploring a range of titles, researchers can identify common patterns, distinctive structures, and the types of language used to convey key themes and contributions. This exposure to various styles can inspire researchers to experiment with different approaches and tailor their titles to best suit their specific research objectives and audience. Furthermore, researchers can use existing literature as a resource to identify gaps in the research landscape that their study can address. Analyzing titles of related studies can reveal areas that have already been extensively explored and those that remain relatively unexplored. This awareness of existing gaps can guide researchers in formulating a research title that highlights the novelty and significance of their study within the broader research context.

Researchers should also consider titles from classic or landmark studies that have made significant contributions to their field. Titles of influential works often have an enduring impact and have stood the test of time. Analyzing these titles can offer insights into the timeless elements of title formulation and how to create a title that encapsulates the enduring relevance and impact of the research. However, while seeking inspiration from existing literature is valuable, researchers should always prioritize originality in their title creation. While it is acceptable to be inspired by existing titles, directly copying or closely imitating the titles of other studies is discouraged. Plagiarism of titles undermines the authenticity and scholarly integrity of the research. Instead, researchers should aim to synthesize their unique research focus, findings, and objectives into a title that distinguishes their study from others while still benefiting from the insights gained from existing literature.

Seeking inspiration from existing literature is a valuable strategy when crafting a research title. Reviewing titles from reputable sources offers researchers insights into effective title construction, disciplinary conventions, and creative possibilities. By drawing inspiration from a diverse range of titles, researchers can gain a nuanced understanding of how to formulate a title that effectively communicates the research's focus, contributions, and appeal. While leveraging existing literature for inspiration, researchers must ensure that their title remains authentic, original, and a true reflection of their unique study's objectives and significance. A well-crafted research title, influenced by the wisdom of existing literature, can act as a captivating and informative introduction to the research, setting the stage for readers to explore the study's valuable insights and scholarly contributions.

2.9 Theoretical Framework or Conceptual Model

Incorporating the theoretical framework or conceptual model guiding the research into the title can add depth and context to the study's identity. The theoretical framework serves as the backbone of the research, providing a systematic and structured understanding of the research problem and guiding the formulation of research questions and hypotheses. By referencing the theoretical foundation in the title, researchers can signal the academic rigor and intellectual depth underpinning their study, inviting readers who share similar theoretical interests to engage with the research.

The inclusion of the theoretical framework or conceptual model in the title helps readers quickly grasp the philosophical underpinnings and broader context in which the study is situated. For instance, a research study investigating leadership styles might incorporate the theoretical framework of transformational leadership, leading to a title like "Exploring the Impact of Transformational Leadership Styles on Employee Engagement." Moreover, referencing the theoretical framework can also spark the interest of researchers and scholars who specialize in the same theoretical perspective. Such researchers are more likely to recognize the title's relevance to their work and view the study as a potential source of valuable insights or corroborating evidence for their research endeavors. However, it is essential to strike a balance when incorporating the theoretical framework into the title. The title should not become excessively laden with theoretical jargon, as this may alienate readers who are not familiar with the specific theoretical framework. Instead, the reference to the theoretical foundation should be clear and concise, providing just enough information to convey the study's theoretical orientation while remaining accessible to a broader audience.

Researchers should also consider whether the theoretical framework or conceptual model is widely recognized and accepted within their field. While it is acceptable to reference lesser-known theories or emerging frameworks, researchers should be prepared to provide additional context or background information in the title or abstract to ensure readers understand the theoretical context of the study. In some cases, a study may draw on multiple theoretical frameworks or perspectives. In such instances, researchers can creatively integrate the key theoretical dimensions into the title, highlighting the study's multidimensional nature. However, researchers should be cautious not to overwhelm the title with complexity, ensuring that it remains concise and informative. Additionally, researchers should bear in mind that the theoretical framework is only one aspect of the research, and the title should not overshadow the study's main empirical focus or research question. While referencing the theoretical foundation can add depth and context to the title, the primary focus should be on conveying the research's core subject matter and objectives.

Referencing the theoretical framework or conceptual model in the research title can enhance the study's identity and scholarly context. By signaling the theoretical underpinnings, researchers can attract readers with similar theoretical interests and establish their study as a valuable contribution to the theoretical literature in the field. However, researchers should ensure that the reference to the theoretical framework is clear and concise, striking a balance between depth and accessibility. A well-crafted research title that incorporates the theoretical foundation serves as an intellectual beacon, inviting readers who share similar theoretical perspectives to engage with the study's ideas and insights.

2.10 Aligning with the Research's Contribution

Perhaps the most crucial aspect of crafting a suitable research title is its alignment with the research's intended contribution to the field. The title should be a clear and concise reflection of the study's core objectives, findings, or theoretical advancements. It serves as a succinct representation of the research's identity, enticing potential readers to delve into the study's content and discover its scholarly contributions. The research title should be carefully designed to showcase the study's relevance and significance within the broader academic landscape. It should answer the question, "What unique value does this research bring to the field?" A title that accurately conveys the research's contribution can position the study as an essential addition to the existing body of knowledge and pique the interest of researchers seeking to stay abreast of the latest developments in the field.

The title should emphasize the research's central theme, innovative methodology, or novel insights. For example, a study introducing a groundbreaking medical intervention might have a title like "A Novel Therapeutic Approach for Alleviating Chronic Pain: Assessing the Efficacy of Neuro-Modulation." Additionally, the research title can highlight the practical implications or potential applications of the study's findings. Titles that emphasize the real-world impact of the research can attract the attention of practitioners, policymakers, and stakeholders who are interested in evidence-based solutions to pressing issues.

Researchers should also consider the target audience when aligning the title with the research's contribution. Different stakeholders may have distinct interests and priorities when selecting research to engage with. For instance, a study investigating the effects of climate change on agricultural productivity might have a title that appeals to both environmental scientists ("Assessing Climate-induced Crop Yield Variability") and policymakers ("Informing Climate-Resilient Agricultural Policies"). Furthermore, the research title can effectively communicate the research's focus on addressing gaps in the existing body of literature. Titles that explicitly reference these gaps and the study's role in filling them can attract readers seeking to explore the latest advancements and unresolved questions in the field.

As the research progresses and findings unfold, researchers should continually review the title's alignment with the study's contributions. If the research takes an unexpected turn or leads to significant discoveries, it might be necessary to revise the title to better reflect the new direction and implications. In some cases, researchers may consider a slightly more generic title to accommodate multiple research outcomes. This approach can be particularly useful in long-term research projects or those with a broader scope, where the precise contribution might only become evident after thorough analysis.

Aligning the research title with the study's contribution is of paramount importance in creating a suitable and impactful title. A well-crafted research title accurately conveys the research's core objectives, findings, or theoretical advancements, inviting readers to explore the study's unique value and relevance to the field. By thoughtfully considering the study's focus, practical implications, target audience, and potential contributions to the existing literature, researchers can create a research title that serves as a compelling introduction to the study's insights and positions it as a significant and valuable addition to the academic discourse.

CHAPTER 3

Writing a Good Introduction to a Research Proposal

ABSTRACT
This chapter delves into the craft of constructing an impactful introduction for a research proposal. It highlights the critical role of an opening statement and hook in capturing readers' attention and establishing the proposal's tone. The importance of background and context formation is emphasized as foundational elements for comprehending the research's origin and relevance. Moreover, the chapter guides the process of identifying the research problem and addressing gaps in the existing literature, laying the groundwork for justifying the study's necessity. It emphasizes clear articulation of research objectives and delineation of the study's scope to guide readers through the research's trajectory. Furthermore, the chapter underscores the significance and justification of the study, elucidating its potential contributions to knowledge and practice. A methodological overview offers insight into the research's approach, preparing readers for subsequent chapters. Ultimately, the chapter provides insights into structuring the proposal to foster reader engagement through a coherent and transparent arrangement.

KEYWORDS: Introduction, opening statement, background, context, research problem, gap identification, justification, methodological overview

The introduction of a research proposal is a critical section that sets the stage for the entire study. It serves as the gateway to the research, capturing the reader's attention, providing essential context, and justifying the need for the study (for details, see Denscombe, 2012; Abdulai & Owusu-Ansah, 2014; Boyack, Smith & Klavans, 2018; Vasanthakumari, 2021). A well-crafted introduction should not only introduce the research topic but also establish its importance, highlight gaps in existing knowledge, and outline the research objectives. In this chapter, we will explore the key elements of writing a good introduction for a research proposal.

3.1 Opening Statement and Hook

The opening statement of the introduction is the research proposal's first impression on the reader, making it a crucial element to capture their attention and create a lasting impact. The primary objective of the opening statement is to engage the reader and compel them to continue reading further into the proposal. To achieve this, researchers can employ various techniques to create a hook that piques the reader's curiosity. One effective way to begin the introduction is by posing a thought-provoking question related to the research topic. The question should be relevant, challenging, and framed in a manner that sparks the reader's interest. For example, in a study on renewable energy sources, the introduction might begin with a question like, "What if we could harness the unlimited power of the sun to meet our energy needs sustainably?"

Another approach is to use a surprising statistic or fact that is relevant to the research topic. Statistics have the power to captivate the reader's attention and emphasize the significance of the research problem. For instance, a proposal on the impact of plastic pollution in the oceans could start with a startling statistic such as, "Every year, over eight million tons of plastic find their way into our oceans, threatening marine life and ecosystems."

Using a compelling anecdote or real-life example can also be an effective way to draw the reader into the research proposal. Anecdotes help humanize the research topic and illustrate its practical implications. For instance, in a proposal investigating the effects of educational interventions, the introduction could begin with a touching story of a student whose life was transformed by a specific teaching approach.

Additionally, researchers can employ a relevant and impactful quote to set the tone for the research proposal. A well-chosen quote from a respected authority or a prominent figure in the field can add credibility and resonance to the research. The quote should be concise, insightful, and directly related to the research problem.

Regardless of the approach used, the opening statement should be concise and to the point. It should immediately convey the research's relevance and importance, giving the reader a clear sense of what to expect from the rest of the proposal. While crafting the opening statement, researchers should keep the target audience in mind and ensure that the hook is appropriate for the intended readers, whether they are academic reviewers, funding agencies, or other stakeholders.

After captivating the reader's attention with the opening statement, researchers should smoothly transition to providing the necessary background information and context for the research. This transition ensures that the reader understands the broader significance of the research problem and sets the stage for a comprehensive and compelling research proposal. By employing a well-crafted hook, researchers can effectively initiate a dialog with the reader, making them eager to explore the research's potential contributions and findings.

3.2 Background and Context

After capturing the reader's attention with the opening statement, the introduction should provide a comprehensive background and context for the research. This section serves as the bridge between the attention-grabbing hook and the core research problem, offering essential information to orient the reader to the study's subject matter.

The background should contextualize the research topic within the broader field of study. It provides a brief overview of the relevant literature, theories, and concepts related to the research problem. By doing so, the background showcases the researcher's understanding of the existing knowledge and highlights the study's continuity with prior research.

Researchers should identify key themes, seminal studies, and major debates that are relevant to the research. A well-structured literature review within the background can help achieve this. This review should be concise yet comprehensive, focusing on the most relevant and influential sources while avoiding excessive details that can be reserved for the full literature review section in the research proposal.

Additionally, researchers should identify gaps or limitations in the existing literature that the proposed study seeks to address. These gaps may relate to unanswered questions, contradictory findings, or unexplored areas within the field. Emphasizing these gaps helps position the research as a logical and timely continuation of the academic discourse, showcasing its originality and contribution to knowledge.

Providing the historical context, if applicable, can also enrich the background. Historical context allows researchers to demonstrate the evolution of the research topic over time, shedding light on how past developments have shaped the current research landscape. It helps establish the continuity and relevance of the proposed study within a broader historical narrative.

Furthermore, the introduction should outline the practical context or real-world relevance of the research. Researchers can discuss how the research problem relates to contemporary issues, challenges, or trends in society or specific industries. By establishing the practical significance of the study, researchers can engage readers who are interested in the practical implications of research findings.

To ensure clarity, researchers should use clear and concise language when presenting the background and context. Technical jargon should be minimized, particularly if the research proposal's audience includes individuals from diverse disciplines or non-experts. The goal is to provide a broad understanding of the research topic that is accessible to a wide range of readers.

As the introduction transitions from the background to the core research problem, the context provided in this section should naturally lead the reader to appreciate the relevance and importance of the study. By presenting the background in a coherent and informative manner, researchers can establish credibility and demonstrate their expertise in the field, setting the stage for a persuasive and compelling research proposal.

3.3 Research Problem and Gap Identification

The research problem is the heart of the research proposal, and the introduction should clearly and concisely articulate it. The research problem is the specific issue or question that the study seeks to investigate and address. It defines the scope and focus of the research, providing a clear sense of direction for the study. When presenting the research problem, it is essential to be precise. Vague or overly broad research problems can lead to confusion and ambiguity in the proposal. Researchers should strive to frame the research problem in a way that is manageable and feasible within the scope of the proposed study. Furthermore, the research problem should be significant and relevant to the field of study. It should address a gap in existing knowledge or contribute to resolving a theoretical or practical issue. This is where the identification of the research gap, which was briefly introduced in the background section, becomes critical.

The research gap refers to the void or deficiency in the current literature that the proposed study intends to fill. It is the space where further investigation is needed to advance understanding, challenge assumptions, or reconcile conflicting findings. Researchers should clearly articulate the specific aspects of the research problem that have not been adequately addressed in previous research. To identify the research gap, researchers can conduct a systematic literature review, analyzing previous studies, and identifying the limitations or unresolved questions. This review helps position the proposed study within the broader scholarly context and demonstrates the necessity and originality of the research.

Researchers should justify the importance of addressing the research gap by explaining its potential implications for the field. This justification serves as the foundation for the research's significance, establishing why the research problem is worth investigating and why the proposed study will contribute valuable insights. Moreover, the introduction should explain how addressing the research problem aligns with the larger objectives of the academic discipline or field of study. Researchers should demonstrate the relevance of the research problem to ongoing scholarly conversations, trends, or emerging areas of interest.

When framing the research problem, researchers should be aware of potential biases and limitations. They should strive for objectivity and impartiality in defining the problem, avoiding preconceived notions or conclusions that could influence the study's findings. Clearly stating the research problem sets the stage for the research's focus, helping readers understand the study's purpose and scope. Additionally, the research problem serves as a basis for formulating research questions, hypotheses, or objectives that guide the study. These elements should be closely aligned with the research problem and reflect the specific outcomes researchers aim to achieve through the study.

The research problem and gap identification are fundamental components of the research proposal introduction. By clearly articulating the research problem, demonstrating its significance, and justifying the need to address the research gap, researchers lay the groundwork for a compelling and well-grounded proposal. A carefully defined research problem ensures that the study remains focused and relevant, positioning it as a valuable contribution to the field and garnering support from readers, reviewers, and potential funders.

3.4 Research Objectives and Scope

Following the research problem, the introduction should present the research objectives or specific aims of the study. The objectives should be clear, measurable, and directly related to addressing the research problem. Moreover, it is crucial to outline the scope of the research by defining the boundaries of the study and specifying what will be included and excluded.

3.5 Significance and Justification

One of the most crucial elements of the introduction is the significance and justification of the research. Researchers must explain why the study is worth undertaking and how it will contribute to the field's knowledge. This can include discussing the potential theoretical, practical, or policy implications of the research findings. Additionally, researchers should highlight the benefits that the study will bring to the academic community, industry, or society at large.

3.6 Methodological Overview

Toward the end of the introduction, a brief methodological overview can be provided. This involves outlining the research design, data collection methods, and data analysis techniques that will be used in the study. While the methodological details will be discussed more comprehensively in the research proposal's subsequent sections, a concise overview in the introduction helps the reader understand the general approach of the study.

3.7 Structure of the Proposal

To conclude the introduction, researchers can provide an overview of the proposal's structure. This involves briefly mentioning the subsequent sections of the proposal, such as the literature review, theoretical framework, research design, data analysis plan, and timeline. This roadmap helps readers anticipate the organization of the proposal and find specific information more efficiently.

A well-written introduction is crucial for a research proposal as it lays the foundation for the entire study. By starting with a compelling opening statement, providing relevant background and context, identifying the research problem and gap, stating clear research objectives, justifying the study's significance, and offering a methodological overview, researchers can create an introduction that captivates the reader, communicates the importance of the research, and sets the stage for the proposal's subsequent sections. A strong and persuasive introduction increases the likelihood of garnering support from reviewers, funders, or other stakeholders and paves the way for a successful research proposal.

3.8 An Example of Writing a Good Introduction to a Research Proposal

Topic: "Navigating Urban Frontiers: Understanding the Social Dynamics of Rural-to-Urban Migration in India"
In the rapidly transforming urban landscape of India, the influx of individuals from rural to urban areas signifies a profound societal shift. This research proposal aims to unravel the intricate social fabric reshaped by rural-to-urban migration, offering a nuanced exploration of the challenges and opportunities inherent in this demographic transition. Contextually, the proposal contextualizes the socio-economic promises and complexities of urban migration in the Indian setting. As the background unfolds, the narrative emphasizes the need for a comprehensive investigation into the varied experiences of migrants within urban environments, recognizing the unique socio-cultural milieu of India.

Identifying a critical gap in the existing literature, the research problem centers on the complex social dynamics that migrants navigate in urban spaces. The research objectives seek to dissect migration patterns, delve into the root causes of social disparities, and propose actionable strategies for fostering social cohesion in the Indian urban landscape.

The research scope extends beyond demographic analysis, encompassing a holistic exploration of social interactions, cultural adaptations, and individual resilience within urban settings. The study's significance lies in its potential to influence urban policies, community development initiatives, and support structures in India. Emphasizing the need for this research, the proposal underscores its capacity to provide empirical insights that can inform interventions promoting inclusivity and social well-being in the dynamic urban spaces of India.

Methodologically, the proposal outlines a mixed-methods approach, employing quantitative surveys and qualitative interviews to capture both demographic trends and subjective experiences of migrants. This methodological choice ensures a comprehensive understanding of the complex research problem. Structurally, the proposal unfolds cohesively, guiding the reader seamlessly from introduction to background, problem identification, objectives, and the study's broader significance in the context of India's evolving social landscape.

CHAPTER 4

Writing the Problem Statement of a Research Proposal

ABSTRACT
This chapter delves into the art of crafting a compelling problem statement within a research proposal. It guides researchers through the process of identifying the research problem and offers techniques for framing it effectively. Emphasis is placed on providing context and background to enable readers to grasp the historical and conceptual underpinnings of the issue at hand. Moreover, the chapter explores the significance of the problem, underscoring its relevance within the larger academic and practical landscape. It navigates the nuanced task of articulating clear research objectives stemming from the identified problem, highlighting the importance of conciseness and precision for presenting a focused and manageable research challenge. Furthermore, readers are advised to avoid presumptive conclusions and instead foster open-ended exploration. The chapter underscores the importance of realism and feasibility in accurately portraying the problem, with transparent acknowledgment of limitations lending credibility to the proposal. Additionally, alignment with research ethics is crucial for maintaining integrity. Ultimately, the chapter emphasizes the need to justify the research's necessity, creating a compelling case for the investigation's urgency and potential impact.

KEYWORDS: Problem statement, context, background, research objectives, feasibility, limitations, research ethics, justification

4.1 Identify the Research Problem

The first and most critical step in writing the problem statement of a research proposal is to identify and articulate the research problem concisely and clearly. The research problem is the central issue or question that the study seeks to investigate and address. It represents the gap or deficiency in knowledge that the research aims to fill. A well-defined research problem not only provides direction and focus to the study but also lays the groundwork for its entire rationale and significance (see Reddy, 2019; Tabatabaei & Tayebi, 2022). To effectively identify the research problem, researchers should engage in a comprehensive and systematic review of the existing literature, theories, and empirical studies related to their field of interest. This review helps researchers identify gaps, controversies, or unresolved questions that form the basis for their research problem. By conducting a thorough literature review, researchers can ensure that their proposed study builds upon existing knowledge and makes a meaningful contribution to the field.

A clear research problem is specific and narrow, avoiding ambiguity or generality. Researchers should avoid broad or vague statements that may encompass multiple issues or lack a distinct focus. Narrowing down the research problem helps ensure that the study remains manageable and feasible within the available resources and time frame. Moreover, the research problem should be relevant and timely. It should align with the current priorities, trends, and challenges in the field of study. A relevant research problem captures the attention of readers and stakeholders, who can appreciate its immediate applicability and potential impact.

To facilitate the identification of the research problem, researchers can use various techniques such as brainstorming, discussing with peers, and seeking advice from mentors or experts in the field. Careful consideration of the research problem from multiple perspectives can lead to a more nuanced understanding and a well-defined problem statement. Additionally, researchers should be aware of the broader context and implications of the research problem. They should consider how addressing the research problem could advance the field, contribute to policy or practice, or benefit society. Articulating the potential impact of the research problem enhances the proposal's persuasive power and underscores the study's significance.

As researchers work on the problem statement, they should continually check its alignment with the research objectives, ensuring that the statement precisely reflects the objectives the study aims to achieve. This alignment ensures that the research problem is the driving force behind the study's methodology, data collection, and data analysis. Lastly, researchers should be receptive to feedback and be willing to refine the problem statement based on the insights and suggestions of colleagues, mentors, or reviewers. Iterative revisions can strengthen the problem statement, making it more focused, coherent, and compelling.

Identifying the research problem is the foundational step in writing the problem statement of a research proposal. A well-defined research problem sets the stage for the entire study, providing a clear sense of direction and significance. Researchers should ensure that the problem statement is specific, relevant, and timely, while also considering its implications and potential impact. By engaging in a systematic review of the literature, aligning the research problem with the objectives, and being open to feedback, researchers can craft a robust and effective problem statement that forms the basis for a successful research proposal.

4.2 Provide Context and Background

Once the research problem has been identified, the problem statement of a research proposal should provide the necessary context and background information to help readers understand the broader significance of the problem and its relevance within the field of study. Contextualizing the research problem involves situating it within the larger academic discipline, research area, or industry. Researchers should offer a concise overview of the subject matter, explaining the key concepts, theories, or themes related to the problem. This helps readers who may not be familiar with the specific research area to grasp the essential context and understand the problem's importance. To provide a comprehensive background, researchers should conduct a thorough literature review of relevant sources related to the research problem. The literature review serves multiple purposes within the problem statement:

(a) Establishing Research Foundation: The literature review demonstrates that the research problem is grounded in existing knowledge and is not redundant or duplicative of previous studies. By highlighting the key studies, theories, and findings related to the problem, researchers showcase the foundation upon which their research will build.

(b) Identifying Gaps and Controversies: Through the literature review, researchers can identify gaps or controversies in the existing knowledge that the proposed study seeks to address. This gap identification clarifies the unique contribution the research will make to the field and justifies its relevance.

(c) Evaluating Prior Research: Researchers can critically evaluate previous research methodologies, data sources, and limitations through the literature review. Understanding the strengths and weaknesses of prior studies helps researchers design a more robust and effective research proposal.

(d) Highlighting Research Trends: The literature review also reveals emerging research trends or areas of interest within the field. Researchers can use this information to position their study within the current research landscape and align it with contemporary research priorities.

Moreover, the background section of the problem statement should provide a concise historical perspective, if applicable. Researchers can explain how the research problem has evolved and how past developments have shaped the current state of the field. This historical context illuminates the continuity of research efforts and illustrates the problem's relevance in the present day. To maintain clarity and conciseness, researchers should focus on the most relevant and significant literature related to the problem. The objective is not to provide an exhaustive review of all studies but to offer a strategic selection that supports the problem statement's rationale. Furthermore, researchers should ensure that the background information is presented in a reader-friendly manner, avoiding excessive technical jargon or complexities. The goal is to provide the necessary context without overwhelming the reader with unnecessary details. By providing a well-structured context and background, researchers demonstrate their understanding of the research area, its historical development, and the gaps in existing knowledge. This prepares the reader to appreciate the significance of the research problem and lays the foundation for the subsequent sections of the research proposal. A comprehensive and well-researched background reinforces the problem statement's credibility and strengthens the proposal's overall persuasiveness.

4.3 Emphasize the Significance

After introducing the research problem and providing the necessary context, the problem statement of a research proposal should emphasize the significance of the identified research problem. This section highlights the reasons why the proposed study is worth conducting and how its findings will contribute to the advancement of knowledge or address practical issues.

Addressing the "So What?" Factor: Emphasizing the significance is essential because it answers the fundamental question of "So what?" Why should the research problem matter to the academic community, stakeholders, or the broader society? Researchers should clearly articulate the potential impact of their study and explain why investing time, resources, and effort in the proposed research is worthwhile.

Contributing to the Academic Discourse: The problem statement should underscore how the research will contribute to the academic discourse and add value to the existing body of knowledge. Researchers should explain how their study will fill a gap in the literature, challenge prevailing theories, or offer novel insights. By positioning the research as a significant contribution to the field, researchers elevate the importance of their work and engage readers who are invested in academic advancements.

Addressing Practical Concerns: In addition to academic significance, researchers should address the practical implications of their research. This involves explaining how the study's findings could have real-world applications, influence policy decisions, or address practical challenges. Whether the research pertains to healthcare, education, technology, or any other field, demonstrating its practical relevance enhances its value and encourages stakeholders to support its execution.

Highlighting the Potential Benefits: Researchers should communicate the potential benefits that may arise from addressing the research problem. This could include improved processes, enhanced outcomes, cost savings, or more efficient interventions. By outlining the potential benefits, researchers demonstrate that their study has tangible and positive outcomes that can justify its execution.

Supporting Broader Goals and Agendas: The problem statement should also align with broader goals, initiatives, or agendas within the academic community or society. Researchers can demonstrate how their research aligns with global challenges, sustainability objectives, or social justice initiatives. Aligning with larger goals enhances the research's credibility and makes it more appealing to readers who seek research that aligns with broader interests.

Addressing Practical Needs: Researchers should consider the needs of stakeholders and address how the proposed research meets those needs. Whether the study is intended to inform policy decisions, address community concerns, or improve organizational practices, researchers should clearly articulate how their research directly addresses those practical needs.

Demonstrating Intellectual Merit: Lastly, the problem statement should highlight the intellectual merit of the proposed research. Researchers should emphasize the novelty, innovation, or unique approach of their study. By showcasing their intellectual merit, researchers position their work as pioneering and deserving of attention and support.

Emphasizing the significance of the research problem is a critical aspect of the problem statement in a research proposal. By addressing the "So what?" factor, contributing to the academic discourse, highlighting practical implications and benefits, supporting broader agendas, addressing stakeholders' needs, and demonstrating intellectual merit, researchers elevate the importance of their study. An emphasis on significance engages readers and stakeholders, compelling them to recognize the research's value and support its execution. By clearly communicating the potential impact and benefits, researchers enhance the proposal's persuasiveness and increase the likelihood of garnering support for their proposed research.

4.4 State the Research Objectives

After introducing the research problem and emphasizing its significance, the problem statement of a research proposal should explicitly state the research objectives. Research objectives are specific, measurable, and achievable goals that the study aims to achieve. They provide a clear roadmap for the research, guiding the entire study's design, methodology, and analysis.

Specificity and Clarity: Research objectives should be formulated in a specific and clear manner. Each objective should address a particular aspect of the research problem and should not be vague or ambiguous. The use of clear language ensures that the research objectives are easily understandable to readers and stakeholders from diverse backgrounds.

Alignment with the Research Problem: The research objectives should be directly aligned with the research problem stated earlier in the proposal. Each objective should serve the purpose of addressing or investigating a specific aspect of the research problem. By ensuring this alignment, researchers demonstrate the logical progression of the study and its relevance to the stated problem.

Measurability and Achievability: Research objectives should be measurable, meaning that their achievement can be assessed through observable and quantifiable outcomes. Measurable objectives allow researchers to gauge the progress and success of the study. Additionally, the objectives should be achievable within the scope of the proposed research, considering the available resources, time constraints, and practical limitations.

Hierarchy and Sequence: In cases where multiple research objectives are stated, they should be presented hierarchically and sequentially. The objectives should be arranged in a logical order, with each objective building upon the previous one. This hierarchical structure helps readers understand the progressive nature of the study and how each objective contributes to the overall research.

Addressing Different Aspects of the Problem: Research objectives should encompass various dimensions of the research problem, ensuring a comprehensive investigation. Each objective should focus on a different aspect or perspective related to the problem. This approach enables researchers to explore the problem from multiple angles, providing a holistic understanding and comprehensive insights.

Guiding the Research Process: The research objectives play a crucial role in guiding the research process. They influence the research design, data collection methods, and data analysis techniques chosen for the study.

By clearly stating the objectives, researchers lay the foundation for the research methodology and demonstrate how each component of the study is purposefully aligned with achieving the stated goals.

Justification of Objectives: Within the problem statement, researchers may briefly justify each research objective. Justification involves explaining why each objective is essential and how its attainment will contribute to addressing the research problem or advancing knowledge. Providing a rationale for each objective adds depth to the problem statement and reinforces its significance.

Interdependence of Objectives: In some cases, research objectives may be interdependent, with the achievement of one objective influencing the attainment of others. Researchers should be aware of these relationships and explain how certain objectives are linked and contribute to a coherent research strategy.

Alignment with Expected Outcomes: The problem statement should also mention how the research objectives are aligned with the expected outcomes of the study. The objectives should be formulated in a way that reflects the anticipated findings and contributes to answering the research questions posed in the proposal.

Clearly stating the research objectives in the problem statement is crucial for the coherence and effectiveness of a research proposal. Specific, measurable, and achievable objectives guide the research process, demonstrate the study's relevance to the research problem, and provide a framework for evaluating the study's success. By ensuring alignment with the research problem, justifying its importance, and presenting it in a logical sequence, researchers create a well-structured and persuasive problem statement that sets the stage for a rigorous and purposeful research proposal.

4.5 Be Concise and Precise

One of the essential qualities of an effective problem statement in a research proposal is conciseness and precision. A concise problem statement communicates the core issue succinctly, without unnecessary elaboration or verbosity. Being precise ensures that the problem is well-defined and leaves no room for ambiguity or misinterpretation. Both qualities contribute to the clarity and impact of the problem statement, making it easier for readers to understand the research problem and its significance.

Avoiding Redundancy and Irrelevance: Conciseness requires researchers to carefully choose their words and avoid unnecessary repetition. Every sentence in the problem statement should serve a distinct purpose and contribute to the overall message. Redundancy can dilute the impact of the problem statement and distract readers from the core issue. Similarly, irrelevant information or tangential details should be omitted to maintain focus and clarity.

Streamlining Language and Syntax: Researchers should aim for clear and straightforward language in the problem statement. Avoiding overly complex sentence structures or technical jargon can help ensure that the statement is easily comprehensible to a broader audience. When using discipline-specific terminology, researchers should provide clear explanations or definitions to aid reader understanding.

Focusing on the Core Issue: A concise problem statement focuses solely on the main research problem without delving into peripheral topics. Researchers should resist the temptation to include background information or tangential details that are not directly relevant to the problem. Staying focused on the core issue helps maintain the statement's clarity and reinforces its significance.

Providing Necessary Context: While being concise, the problem statement should still offer the necessary context and information to enable readers to grasp the research problem's importance. This context-setting ensures that the statement does not become cryptic or disconnected from the broader research area. Balancing conciseness with contextual information requires researchers to carefully select the most relevant and impactful details.

Using Quantitative and Qualitative Measures: Precision in a problem statement involves the use of specific quantitative or qualitative measures to describe the research problem. This could include quantitative data such as percentages, numbers, or statistics to quantify the magnitude of the problem. For qualitative research, researchers may use descriptive language or case examples to illustrate the problem's impact. Providing specific measures helps readers understand the problem's scale and importance.

Avoiding Ambiguity and Generalization: A precise problem statement leaves no room for ambiguity or misinterpretation. Researchers should use unambiguous language to express the research problem and avoid vague or overly general statements. Ambiguity can lead to confusion or multiple interpretations, hindering the proposal's clarity and impact.

Iterative Refinement: Achieving conciseness and precision often requires multiple rounds of revision and refinement. Researchers should continually review and edit the problem statement to ensure that it conveys the essential message concisely and precisely. Seeking feedback from peers, mentors, or colleagues can be beneficial in identifying areas for improvement.

Clarity for Different Audiences: Researchers should consider the proposal's diverse audience, which may include experts, funding agencies, institutional reviewers, and non-specialists. Striking the right balance between conciseness and precision ensures that the problem statement remains accessible and understandable to readers with different levels of expertise.

Conciseness and precision are crucial attributes of an effective problem statement in a research proposal. By avoiding redundancy and irrelevance, streamlining language, focusing on the core issue, providing necessary context, using specific measures, avoiding ambiguity and generalization, and iteratively refining the statement, researchers can craft a problem statement that is clear, impactful, and compelling. A well-constructed problem statement captures the essence of the research problem succinctly, drawing readers' attention and compelling them to recognize the proposal's significance.

4.6 Avoid Presumptive Conclusions

In the problem statement of a research proposal, it is crucial to avoid making presumptive conclusions or biases that could undermine the proposal's objectivity and credibility. A problem statement should remain impartial and neutral, focusing solely on describing the research problem without preconceived notions about the study's outcomes.

Objective and Unbiased Language: Researchers should use objective and unbiased language when formulating the problem statement. They should avoid incorporating personal opinions, preferences, or value judgments that could introduce bias. Instead, researchers should present the problem based on existing evidence and the identified research gap.

Avoiding Loaded Language: Loaded language, which contains emotionally charged words or phrases, should be avoided in the problem statement. Such language can influence readers' perceptions and create a bias toward a particular stance or interpretation. By using neutral and factual language, researchers maintain the problem statement's integrity and encourage readers to objectively evaluate the research proposal.

Balanced Presentation of Information: A well-constructed problem statement presents a balanced view of the research problem. Researchers should acknowledge and discuss different perspectives or competing theories, if applicable, without favoring one over the other. By acknowledging diverse viewpoints, researchers demonstrate their intellectual rigor and openness to exploring multiple aspects of the problem.

Avoiding Overgeneralization: Researchers should be cautious not to overgeneralize the research problem based on limited evidence or specific cases. Overgeneralization can lead to flawed conclusions and misrepresentation of the problem's true scope. Instead, the problem statement should focus on a specific, well-defined research problem supported by relevant evidence.

Relying on Empirical Evidence: To avoid presumptive conclusions, researchers should ground the problem statement in empirical evidence and established literature. Conclusions or assumptions based solely on personal beliefs or anecdotal experiences lack scientific validity. Instead, researchers should draw on reputable sources and empirical studies to support their problem statement.

Recognizing Limitations of Existing Research: In formulating the problem statement, researchers should recognize the limitations of existing research and avoid overstating the potential impact of their study. Acknowledging the current gaps and constraints in the literature positions the proposed research as a valuable contribution rather than an attempt to solve all related issues.

Remaining Open to Alternative Findings: Researchers should maintain an open mind and remain receptive to alternative findings or unexpected results in their study. The problem statement should not predetermine the outcomes; rather, it should be designed to investigate and explore the research problem systematically and rigorously.

Peer Review and Feedback: Engaging in peer review and seeking feedback from colleagues, mentors, or research partners can help identify any potential presumptive conclusions in the problem statement. Reviewers can offer valuable insights and challenge researchers to refine their statements, ensuring their objectivity and neutrality.

Adhering to Research Ethics: Finally, adhering to research ethics is paramount in the problem statement. Researchers should conduct the study with integrity, honesty, and transparency. The problem statement should not be a means to justify biased research or support predetermined outcomes.

Avoiding presumptive conclusions in the problem statement of a research proposal is essential for maintaining the proposal's credibility and objectivity. By using objective language, avoiding loaded terminology, presenting balanced information, relying on empirical evidence, recognizing limitations of existing research, remaining open to alternative findings, seeking peer review and feedback, and adhering to research ethics, researchers create a problem statement that is impartial, rigorous, and trustworthy. A problem statement free from presumptive conclusions fosters confidence in the research proposal and positions the study as an unbiased investigation of a well-defined research problem.

4.7 Be Realistic and Feasible

In the problem statement of a research proposal, it is crucial to present a research problem that is realistic and feasible within the available resources, time constraints, and research scope. A realistic problem statement ensures that the proposed research is practical and can be successfully executed, while a feasible problem statement demonstrates that the study is achievable and aligns with the researcher's capabilities.

The Practicality of Research Design: Researchers should consider the practicality of the research design when formulating the problem statement. This involves assessing whether the proposed methodology, data collection methods, and analysis techniques are viable and appropriate for addressing the research problem. A well-designed study should be logistically achievable and produce meaningful results.

Available Resources: The problem statement should align with the resources available to the researchers. This includes financial resources, access to data or participants, laboratory equipment, and other necessary facilities. A realistic problem statement takes into account the funding and support needed for the research and outlines how these resources will be obtained.

Time Constraints: Time is a critical factor in research projects. Researchers should ensure that the problem statement is feasible within the proposed timeline. They should assess whether the study can be completed within the allocated time frame and whether the research objectives are achievable within that period.

Avoiding Overambitious Goals: While ambitious research projects can be admirable, the problem statement should avoid overambitious goals that might be beyond the researcher's capacity. Unrealistic objectives can lead to disappointment and potential failure to meet research expectations. Instead, researchers should focus on meaningful and achievable objectives that align with the available resources and expertise.

Considering Ethical Constraints: Feasibility also involves considering ethical constraints related to the research. Researchers must ensure that the study design and procedures adhere to ethical guidelines and protect the rights and well-being of human participants, animal subjects, or the environment, as applicable.

Pilot Studies and Pretesting: Before finalizing the problem statement, researchers may conduct pilot studies or pretesting to assess the feasibility of the proposed research. Pilot studies allow researchers to identify potential challenges, refine methodologies, and make necessary adjustments before implementing the full study.

Flexibility and Adaptability: While the problem statement should be realistic, researchers should also build flexibility and adaptability into their research plans. Unforeseen challenges or changes may arise during the study, and researchers should be prepared to modify their approach accordingly.

Anticipating Potential Risks: Realistic and feasible problem statements acknowledge potential risks and challenges that could impact the research. By anticipating these risks, researchers can develop contingency plans to address unexpected issues and maintain the study's progress.

Impactful within Limitations: A well-crafted problem statement demonstrates that the proposed research, while realistic and feasible, can still have a meaningful impact within its limitations. Researchers should clarify how the study's findings can contribute to knowledge, address the research problem, or have practical implications, even if the study is not intended to be exhaustive.

Being realistic and feasible in the problem statement of a research proposal is vital for ensuring that the proposed study can be successfully executed and produce valuable results. By considering the practicality of the research design, available resources, time constraints, ethical considerations, and potential risks, researchers demonstrate a thoughtful and well-planned approach. Balancing ambition with practicality, the problem statement sets the stage for a research proposal that is credible, achievable, and positioned for success.

4.8 Be Transparent about Limitations

In the problem statement of a research proposal, it is essential to be transparent about the limitations that may impact the proposed study. Acknowledging the potential limitations upfront demonstrates the researchers' awareness of the challenges and constraints that could affect the research and reinforces the proposal's credibility.

Identifying Potential Limitations: Researchers should critically assess their proposed research and identify potential limitations that could arise during the study. Limitations can stem from various factors, including the research design, data collection methods, sample size, data analysis techniques, time constraints, and access to resources or participants. By carefully considering each aspect of the study, researchers can anticipate potential shortcomings.

Impact on Research Validity: Transparency about limitations is crucial for maintaining research validity. Readers and reviewers can evaluate the potential impact of these limitations on the study's findings and conclusions. Researchers should be candid about how certain limitations may affect the study's reliability and the extent to which the results can be generalized to a broader population or context.

Mitigation Strategies: While identifying limitations is essential, researchers should also propose strategies to mitigate their impact. Mitigation strategies may include refining the research design, enhancing data collection methods, addressing potential biases, expanding the sample size, or using alternative analysis techniques. Demonstrating that the researchers have considered ways to minimize the limitations strengthens the proposal's robustness.

Limitations vs. Delimitations: Researchers should distinguish between limitations and delimitations in the problem statement. Limitations refer to factors that could potentially impact the research's internal or external validity, while delimitations are intentional choices made by the researchers to narrow the scope of the study. Delimitations help focus the research on specific variables, populations, or contexts and are not considered weaknesses in the study.

Ethical Considerations as Limitations: Ethical considerations may also be acknowledged as potential limitations. For example, if the research involves human participants, researchers should be transparent about potential challenges in obtaining informed consent, maintaining confidentiality, or addressing sensitive topics. Ethical limitations demonstrate the researchers' commitment to conducting the study ethically and responsibly.

Transparency Builds Trust: Being transparent about limitations is a sign of scholarly integrity and builds trust with readers and reviewers. Transparency indicates that the researchers are forthcoming about the study's potential shortcomings and are committed to conducting rigorous and honest research. This openness enhances the proposal's credibility and supports the case for funding or approval.

Learning Opportunities: Researchers can view limitations as learning opportunities and avenues for future research. Transparently discussing limitations allows researchers to reflect on how they might refine their study in the future or explore related research questions. Learning from limitations can contribute to the growth of knowledge in the field.

Honest Communication: In the problem statement, researchers should communicate limitations honestly and straightforwardly. While it may be tempting to downplay limitations, transparent communication allows readers to make informed judgments about the research's strengths and weaknesses.

Being transparent about limitations in the problem statement is vital for a comprehensive and credible research proposal. By identifying potential limitations, discussing their impact on research validity, proposing mitigation strategies, distinguishing limitations from delimitations, addressing ethical considerations, building trust through transparency, viewing limitations as learning opportunities, and communicating with honesty, researchers create a problem statement that is realistic and candid. Acknowledging limitations upfront positions the research proposal as a well-considered and responsible endeavor, enhancing its chances of receiving support and approval for the proposed study.

4.9 Align with Research Ethics

In the problem statement of a research proposal, it is imperative to ensure that the research aligns with ethical principles and guidelines. Ethical considerations are essential to safeguard the rights, well-being, and dignity of research participants, as well as the integrity of the research process itself. Demonstrating a commitment to ethical research practices enhances the proposal's credibility and demonstrates the researchers' responsibility towards their participants and the broader scientific community.

Respect for Participants' Autonomy: Researchers should emphasize the importance of respecting participants' autonomy in the problem statement. This involves obtaining informed consent from participants, ensuring they have the right to withdraw from the study at any time, and protecting their privacy and confidentiality. Researchers should clearly articulate how they will uphold these principles throughout the research process.

Beneficence and Non-Maleficence: Beneficence refers to the researchers' obligation to maximize benefits and minimize potential harm to participants. Non-maleficence involves avoiding causing harm to participants intentionally. In the problem statement, researchers should outline how they will uphold these ethical principles and mitigate any risks that may arise during the study.

Avoiding Coercion and Exploitation: Researchers must ensure that participants are not coerced or unduly influenced to participate in the study. Additionally, researchers should be mindful of power dynamics and avoid any exploitation of vulnerable populations. The problem statement should address how researchers plan to mitigate these risks and create a safe research environment for all participants.

Balancing Risks and Benefits: Ethical research involves carefully balancing the risks and benefits associated with the study. In the problem statement, researchers should discuss the potential risks participants may face and the measures in place to minimize those risks. They should also highlight the potential benefits of the research to participants and society as a whole.

Research with Special Populations: If the research involves special populations, such as children, elderly individuals, individuals with disabilities, or marginalized communities, researchers should address the unique ethical considerations associated with these groups. The problem statement should demonstrate how researchers will ensure their special needs and vulnerabilities are taken into account.

Compliance with Institutional Review Boards (IRBs): Researchers should indicate in the problem statement that the proposed research will undergo review and approval by an Institutional Review Board or Ethics Committee. IRBs play a crucial role in assessing the ethical implications of research studies and ensuring that they adhere to ethical standards.

Data Handling and Privacy Protection: Ethical data handling is a critical aspect of research. Researchers should discuss in the problem statement how they will collect, store, and protect participants' data to ensure confidentiality and privacy. Compliance with data protection regulations and guidelines should be emphasized.

Cultural Sensitivity and Respect for Diversity: Researchers should be culturally sensitive and respectful of participants' diverse backgrounds and beliefs. The problem statement should describe how researchers will account for cultural differences and ensure that the research is conducted in a culturally appropriate manner.

Transparency in Reporting: Ethical research requires transparent reporting of the study's methods, results, and interpretations. In the problem statement, researchers should commit to honest and accurate reporting, acknowledging any potential biases or limitations that may arise during the study.

Aligning with research ethics is paramount in the problem statement of a research proposal. By emphasizing respect for participants' autonomy, beneficence, non-maleficence, avoiding coercion and exploitation, balancing risks and benefits, addressing special populations, complying with IRBs, handling data and privacy ethically, being culturally sensitive, and committing to transparent reporting, researchers demonstrate their commitment to conducting responsible and ethical research. Ethical considerations are essential not only for the protection of research participants but also for upholding the integrity and trustworthiness of the research itself.

4.10 Justify the Need for the Research

In the problem statement of a research proposal, it is crucial to provide a compelling justification for the need to undertake the proposed study. Justification serves as the backbone of the research proposal, explaining why the research problem is worth investigating and how the study will contribute to knowledge, address gaps in existing literature, or provide practical solutions to real-world challenges.

Addressing Knowledge Gaps: One of the primary justifications for research is addressing knowledge gaps in the existing literature. The problem statement should demonstrate that the proposed study will build upon previous research, extend current knowledge, or challenge prevailing assumptions. By presenting a thorough literature review, researchers can identify the gaps in the field and articulate how their study will fill those gaps.

Relevance and Timeliness: The problem statement should emphasize the relevance and timeliness of the research. Researchers should highlight how the proposed study aligns with current issues, trends, or emerging research areas within the discipline. Demonstrating the timeliness of the research enhances its importance and highlights the potential impact of the findings.

Contributing to Theory and Practice: Justification can be found in the study's potential contribution to both theoretical understanding and practical applications. Researchers should explain how their study will advance existing theories, concepts, or models. Simultaneously, they should discuss how the findings can be translated into practical applications that address real-world problems or inform policy decisions.

Supporting Existing Research or Challenging Assumptions: The problem statement may justify the need for research by supporting and confirming previous findings in the field. Alternatively, researchers may argue that their study aims to challenge established assumptions or theories, providing a fresh perspective on the research problem.

Social or Economic Impact: Researchers should explore the potential social or economic impact of their study. This involves discussing how the research findings could lead to positive changes in society, improve quality of life, or contribute to economic growth. Such discussions are particularly relevant for research proposals seeking funding from governmental or non-governmental agencies.

Academic and Intellectual Significance: The justification should encompass the academic and intellectual significance of the research. Researchers should discuss how the study will contribute to the academic discourse, expand the field's knowledge base, or open new avenues of research. Emphasizing the intellectual merit of the research enhances its scholarly value.

Addressing Practical Needs: A strong justification also considers the practical needs that the research aims to address. This could involve solving problems faced by industries, communities, or specific user groups. Researchers should articulate how their study's outcomes will have practical implications and potential applications.

Building on Researcher Expertise: The problem statement can justify the need for research by highlighting the researcher's expertise in the field. If the study aligns with the researcher's past work or area of specialization, this expertise can enhance the proposal's credibility and the potential for high-quality research.

Stakeholder Support and Collaboration: Researchers may demonstrate the need for research by securing support from relevant stakeholders, organizations, or potential collaborators. Stakeholder involvement can validate the importance of the research problem and its potential impact on different sectors of society.

4.11 An Example of the Problem Statement of a Research Proposal

Title: "Addressing Educational Disparities in Digital Learning Environments in Mexico"

In the dynamic landscape of Mexico's digital education transition, a critical issue surfaces—the persistent educational disparities among students. As the country embraces rapid digitization in its educational systems, there is a growing need to dissect the underlying factors contributing to uneven learning outcomes. The introduction of technology in education, while promising greater access and flexibility, has inadvertently accentuated disparities based on socioeconomic factors, digital access, and technological proficiency. This study is prompted by the imperative to comprehensively understand these contextual nuances, recognizing the complexity of the challenges faced by students in the digital learning environment.

The educational disparities identified in this research extend beyond mere academic performance, delving into the multifaceted dimensions of digital education. Socio-economic disparities, varying levels of digital access, and divergent technological competencies all contribute to a complex web of challenges. This study recognizes the intricate interplay of these factors and aims to untangle the threads that weave into educational disparities. By avoiding presumptive conclusions, the research maintains an objective stance, acknowledging the intricate nature of the issues at hand. Realism and feasibility are at the forefront, ensuring a pragmatic exploration of the scope and impact of the research.

Justifying the need for this research is the urgent call to create inclusive digital learning environments that cater to the diverse needs of Mexican students. The study aspires to provide actionable insights for educators, policymakers, and stakeholders invested in fostering equal opportunities in the digital education landscape. With Mexico's education system undergoing significant transformations, addressing these disparities becomes imperative for building a robust, equitable foundation for the future. This research endeavors not only to diagnose the problem but also to offer practical solutions that can pave the way for a more inclusive and accessible digital education experience in Mexico.

Problem statement is a fundamental aspect of a research proposal. By addressing knowledge gaps, highlighting relevance and timeliness, contributing to theory and practice, supporting existing research or challenging assumptions, assessing social or economic impact, emphasizing academic and intellectual significance, addressing practical needs, building on researcher expertise, and securing stakeholder support, researchers create a persuasive case for the research's significance and potential contributions. A well-justified problem statement increases the proposal's chances of receiving support, funding, or approval for conducting the proposed study.

CHAPTER 5

Literature Review: Conceptual and Theoretical Framework of a Research Proposal

ABSTRACT
In this chapter, a thorough examination of the roles of literature review, conceptual framework, and theoretical framework within a research proposal is presented. It meticulously explores the nuanced distinctions between these elements and their significance in shaping the study. The purpose of a literature review is scrutinized, emphasizing its crucial role in situating the research within existing scholarly conversations. Moreover, the chapter delves into strategies for defining key concepts in the proposal, distinguishing between theoretical and operational definitions to ensure clarity and precision. It navigates the intricate process of crafting robust conceptual and theoretical frameworks, offering practical insights into their construction. Furthermore, through a systematic approach, researchers are guided in aligning these frameworks with the research's objectives and research questions, thereby enhancing the overall coherence and depth of the proposal.

KEYWORDS: Literature review, conceptual framework, theoretical framework, research proposal

5.1 Meanings of Literature Review, Conceptual and Theoretical Framework

The literature review is a critical component of any research proposal or study, providing a comprehensive overview of existing knowledge and research related to the proposed research problem. Within the literature review, a conceptual framework plays a central role in organizing and synthesizing the existing literature (see Wolery et al., 2014; Denscombe, 2012). It serves as a theoretical foundation that guides the study's design, methodology, and interpretation of findings. The conceptual framework helps researchers understand the relationships between variables, concepts, and theories, providing a coherent structure for the research. In a research proposal, a conceptual framework refers to a theoretical structure or model that guides the study's design, methodology, and interpretation of findings. It serves as a foundational framework that helps researchers organize and integrate relevant theories, concepts, and variables related to the research problem. The conceptual framework provides a clear and coherent structure for understanding the relationships between various elements of the study, allowing researchers to develop hypotheses, research questions, and data collection strategies based on established theories and prior research. On the other hand, the theoretical framework in a research proposal is a structured foundation that draws on existing theories and concepts to guide the study's design and interpretation. It contextualizes the research problem within established knowledge, informs research questions and hypotheses, influences methodological choices, and contributes to the overall validity and reliability of the study by aligning it with relevant theoretical perspectives.

5.2 Purpose of Literature Review

The purpose of a conceptual framework in a research proposal is to:

- Theoretical Foundation: Establish a theoretical foundation for the study by drawing on existing theories and models from the literature. The conceptual framework identifies key theories relevant to the research problem and helps researchers apply these theories to their specific study.
- Conceptual Clarity: Provide conceptual clarity by defining and operationalizing the main concepts and variables involved in the study. By clearly defining these elements, the researchers ensure precision and consistency in their investigation.
- Guidance for Research Questions/Hypotheses: Guide the formulation of research questions or hypotheses based on the relationships

between variables proposed in the conceptual framework. The framework allows researchers to make informed predictions about the expected outcomes of the study.

Integration of Literature Findings: Integrate the findings of the literature review into a coherent structure. The conceptual framework helps researchers synthesize the existing knowledge, identify gaps in the literature, and build on previous research.

Framework for Data Analysis: Provide a framework for data analysis by specifying how the collected data will be analyzed to address the research questions or test the hypotheses. It ensures that the data analysis aligns with the theoretical underpinnings of the study.

Contextualization of Results: Contextualize and interpret the study's results within the broader theoretical context. The conceptual framework helps researchers understand the implications of their findings with the existing body of knowledge.

Coherence and Consistency: Ensure coherence and consistency in the research design, as all aspects of the study are guided by the same set of theoretical principles. This enhances the study's internal validity and credibility.

Theoretical Contributions: Facilitate the identification of potential theoretical contributions of the study. Researchers can assess how their findings contribute to or challenge existing theories, adding to the advancement of knowledge in the field.

The conceptual framework is typically presented in the literature review section of the research proposal. It may be depicted as a diagram or model that illustrates the relationships between the main concepts and variables. However, in some cases, it may be presented in a narrative form, outlining the key elements and their connections.

Overall, a well-developed conceptual framework in a research proposal demonstrates the researchers' theoretical understanding, analytical approach, and alignment with the existing body of knowledge. It is a crucial component that underpins the research proposal's rationale and justifies the study's importance and potential contributions to the field.

5.3 How We Can Define the Concepts: Theoretical and Operational Definitions

In a research proposal, defining concepts is a crucial step that involves providing clear and precise explanations of the terms used in the study. There are two main types of definitions: theoretical definitions and operational definitions.

5.3.1 Theoretical Definitions

Theoretical definitions refer to the conceptual understanding of a term based on established theories, existing literature, and the researchers' conceptual framework. These definitions provide a broad and abstract understanding of the concepts under investigation. Theoretical definitions are typically derived from academic sources and scholarly literature. When defining concepts theoretically in a research proposal:

- Review Existing Literature: Conduct a thorough review of relevant academic literature and identify how other researchers have defined and used the concepts in the field. Cite reputable sources to support the theoretical definitions.
- Refer to Theoretical Framework: If the research proposal includes a conceptual framework, draw upon the theories and models in the framework to define the concepts. Explain how the concepts are interrelated and how they fit within the broader theoretical context.
- Use Clear Language: Clearly articulate the theoretical definitions in the proposal using precise and unambiguous language. Avoid jargon or technical terms that may be unclear to the readers.
- Avoid Circular Definitions: Ensure that the theoretical definitions do not rely on the same concept being defined. Avoid circular reasoning, where the definition refers back to the concept itself.

5.3.2 Operational Definitions

Operational definitions, on the other hand, refer to the specific and measurable way in which a concept will be observed, measured, or manipulated in the research study. Operational definitions bridge the gap between abstract theoretical concepts and concrete empirical observations. When defining concepts operationally in a research proposal:

- Specify Measurement Procedures: Describe the specific methods or instruments that will be used to measure or observe the concepts in the study. This could include questionnaires, interviews, observations, or physiological measures.
- Quantify Variables: If the concepts are quantitative variables, specify the units of measurement and the scale used (e.g. Likert scale and interval scale).
- Define Categorical Variables: For categorical variables, clarify the categories and how they will be classified (e.g. gender categories and age groups).
- Explain Manipulation: If the study involves experimental manipulation, describe how the independent variables will be manipulated to assess their impact on the dependent variables.
- Relate to Research Questions/Hypotheses: Ensure that the operational definitions are aligned with the research questions or hypotheses, allowing the study to address the intended objectives.
- Validity and Reliability: Discuss how the operational definitions ensure the validity and reliability of the data collected. Validity refers to the extent to which a measurement accurately captures the concept, while reliability relates to the consistency of the measurement.

By providing both theoretical and operational definitions in a research proposal, researchers demonstrate a comprehensive understanding of the concepts under study and lay the foundation for a well-designed and rigorously conducted research project. The definitions help readers, reviewers, and potential funders grasp the study's focus and significance, ensuring clarity and precision in the research endeavor.

5.4 How Can We Write a Good Conceptual and Theoretical Framework?

Writing a good conceptual framework in a research proposal is essential for providing a strong theoretical foundation for the study. A well-constructed conceptual framework guides the research design, methodology, and analysis, and demonstrates the coherence and rigor of the study. Here are some steps to write a good conceptual framework in a research proposal:

- Identify Key Concepts and Variables: Start by identifying the main concepts and variables that are central to your research problem. These are the key elements that the study will investigate. Clearly define these concepts in both theoretical and operational terms, as discussed in the previous response.

- Conduct a Thorough Literature Review: Conduct a comprehensive literature review to identify existing theories, models, and frameworks relevant to your research problem. Analyze how other researchers have approached similar topics and how they have conceptualized the relationships between different variables. This review will inform the development of your conceptual framework.

- Select Theoretical Perspectives: Choose one or more theoretical perspectives that align with your research problem and objectives. These theoretical perspectives should provide a theoretical lens through which you can understand and interpret your study's findings. The selected theories should be well-established and widely accepted in the field.

- Build Connections between Concepts: Establish clear connections between the key concepts and variables identified in your study. Explain how these concepts relate to one another based on the selected theoretical perspectives. Use the literature review to support these connections and demonstrate how your study builds on existing knowledge.

- Develop a Conceptual Diagram (Optional): Consider creating a visual representation of your conceptual framework using a diagram or model. A conceptual diagram can help illustrate the relationships between the concepts and variables, making it easier for readers to understand the theoretical structure of your study.
- Formulate Research Questions or Hypotheses: Based on the conceptual framework, formulate specific research questions or hypotheses that will guide your study. These research questions or hypotheses should reflect the relationships between the key concepts and variables as proposed in your conceptual framework.
- Ensure Consistency and Coherence: Ensure that the conceptual framework is consistent with the research questions, hypotheses, and overall research objectives. The framework should provide a coherent structure that justifies the research problem and aligns with the proposed research design.
- Link the Framework to the Research Proposal's Introduction: Connect the conceptual framework to the introduction of your research proposal. Clearly explain how the conceptual framework informs the rationale for your study and why it is essential for addressing the research problem.
- Justify Your Choices: Incorporate a section in your research proposal that justifies the selection of specific theories and the conceptual framework. Explain why these choices are the most appropriate and suitable for your research problem.
- Revise and Refine: After writing the initial conceptual framework, seek feedback from colleagues, mentors, or advisors. Revise and refine the framework based on their input, ensuring that it is clear, well-supported, and logically sound.

By following these steps, you can create a strong and well-structured conceptual framework that enhances the quality and impact of your research proposal. A well-developed conceptual framework demonstrates your scholarly rigor, theoretical understanding, and ability to contribute valuable insights to the field of study.

CHAPTER 6

Writing Research Objectives and Research Questions of a Research Proposal

ABSTRACT
This chapter provides a nuanced exploration of the process of formulating research objectives and research questions within a research proposal. It begins by elucidating the essence of research objectives and their pivotal role in framing the trajectory of the study. Distinctions between general and specific objectives are illuminated, emphasizing their collaborative contribution to delineating the focus and direction of the research. Moreover, practical guidance is offered for crafting effective general and specific research objectives that align with the proposal's overarching goals. The chapter then delves into the significance of research questions as pivotal tools for guiding the research process. Types of research questions are outlined, showcasing their diverse roles in shaping research design. Furthermore, the chapter underscores the importance of research questions in driving inquiry and exploration. Researchers are guided in the art of constructing impactful research questions that reflect the intent of the research and resonate with readers, thereby enhancing the clarity and depth of the proposal.

KEYWORDS: Research objectives, general objectives, specific objectives, research questions, impactful construction.

Writing research objectives and research questions is a fundamental step in developing a well-structured and focused research proposal. Clear and concise research objectives and questions set the direction for the study, guiding the research design and data collection. They define the specific aims and scope of the research, providing a roadmap for the entire research process. In this section, we will outline the key principles for crafting effective research objectives and research questions in a research proposal.

6.1 What Is the Meaning of Research Objectives?

Research objectives in a research proposal refer to the specific and measurable goals or aims that the researcher intends to achieve through the study. These objectives outline what the researcher wants to accomplish, what questions they seek to answer, or what problems they aim to solve during the research process. Research objectives provide clarity and focus to the study, guiding the researcher in their data collection, analysis, and interpretation of results (see Smith, Mykhalovskiym & Weatherbee, 2006; Abdulai & Owusu-Ansah, 2014).

The research objectives should be formulated in a clear, concise, and unambiguous manner. They should be closely aligned with the research problem stated in the proposal and should be achievable within the scope and limitations of the study. Each objective should be specific, measurable, attainable, relevant, and time-bound (SMART). This ensures that the objectives are realistic and can be effectively evaluated.

Research objectives typically stem from the research questions or hypotheses and are informed by the theoretical framework or existing literature. They help to narrow down the research focus and provide a roadmap for conducting the study. By defining clear research objectives, the researcher demonstrates a clear purpose and direction for the research, increasing the proposal's credibility and the likelihood of conducting a successful study.

6.2 General Objectives and Specific Objectives

In a research proposal or study, the general objective and specific objectives are two distinct types of research objectives that outline the overall purpose of the research and the specific aims or tasks to be accomplished, respectively. Let's explore each type:

General Objective: The general objective is the broad and overarching goal of the research. It represents the main purpose of the study and provides an overall direction for the research. The general objective answers the question, "What is the main aim of the research?" It is usually formulated as a broad statement that encapsulates the desired outcome or the ultimate impact of the study. The general objective is often placed at the beginning of the research proposal or introduction section. It sets the tone for the entire study and helps the reader understand the research's significance and relevance. While the general objective is broad, it should still be specific enough to reflect the study's main focus and align with the research problem or topic.

Example of a General Objective:
- "To investigate the factors influencing consumer preferences for eco-friendly products in the market."

Specific Objectives: In contrast to the general objective, specific objectives are precise, measurable, and concrete tasks or sub-goals that the researcher aims to achieve to fulfill the general objective. They break down the main goal into smaller, manageable components. Each specific objective represents a distinct aspect of the research that needs to be addressed to reach the overall research goal. Specific objectives are usually listed as bullet points or numbered items under the general objective in the research proposal. They serve as a roadmap for the research, guiding the researcher's actions and providing clarity on what needs to be accomplished.

Example of Specific Objectives:

- To review the literature on consumer behavior and eco-friendly products.
- To identify the key factors influencing consumers' purchase decisions for eco-friendly products.
- To analyze the awareness and perception of eco-friendly products among different consumer segments.

- To assess the impact of eco-labeling and environmental certifications on consumer preferences.
- To propose recommendations for businesses and policymakers to promote the adoption of eco-friendly products.

In summary, the general objective represents the overarching goal of the research, while the specific objectives outline the specific tasks or sub-goals that need to be accomplished to achieve the general objective. Together, they provide a clear and structured framework for the research, ensuring that the study is well-focused, purposeful, and achievable.

6.3 How Can We Write Effective General and Specific Research Objectives?

Writing effective general and specific research objectives in a research proposal requires careful planning, clarity of purpose, and alignment with the research problem. Here are some tips to craft effective general and specific research objectives:

- Understand the Research Problem: Before formulating research objectives, thoroughly understand the research problem or topic. Clearly define the scope of the study and identify the key aspects you want to investigate. A well-defined research problem serves as the foundation for developing objectives that are relevant and meaningful.
- Start with the General Objective: Begin by crafting a clear and concise general objective that states the main aim of the research. The general objective should be specific enough to capture the research focus and broad enough to encompass the study's ultimate goal. Avoid ambiguous or vague language and ensure the general objective aligns with the research problem.

Use Action-Oriented Language: Write research objectives using action-oriented language that conveys what tasks or activities you plan to undertake. Use strong verbs such as "investigate," "examine," "identify," "analyze," "compare," "assess," or "propose." Action-oriented language makes the objectives more explicit and measurable.

Be Specific and Measurable: For specific objectives, be precise about what you want to achieve. Each specific objective should represent a discrete and measurable task or outcome. Use quantitative or qualitative indicators to measure progress or success in achieving specific objectives.

Ensure Feasibility and Realism: Ensure that the research objectives are feasible and realistic within the available resources, time frame, and expertise. Avoid setting objectives that are too ambitious or unattainable. Consider the practical constraints and limitations of the study.

Consider the Research Design: Tailor the specific objectives to fit the chosen research design. For instance, if conducting an experimental study, the objectives may focus on testing causal relationships between variables. If conducting a qualitative study, the objectives may aim to explore experiences or perceptions.

Link Specific Objectives to the General Objective: Ensure that each specific objective is directly linked to the general objective. Each specific objective should contribute to achieving the overall aim of the research. This alignment ensures coherence and consistency in the research objectives.

Use a Logical Sequence: Organize the specific objectives in a logical sequence that reflects the flow of the research. Start with introductory objectives that lay the groundwork for the study and progress to more complex objectives as the study advances.

Avoid Overloading with Objectives: Avoid including too many objectives, as this may lead to a lack of focus and clarity. Prioritize the most critical objectives that directly address the research problem and align with the available resources.

Seek Feedback and Review: Share the research proposal with colleagues, mentors, or advisors to get feedback on the research objectives. Review the objectives for coherence, relevance, and clarity. Revise and refine the objectives based on the feedback received.

By following these tips, you can craft effective and well-structured general and specific research objectives that provide a clear roadmap for your research proposal. Strong objectives demonstrate the significance of your study, outline the tasks to be accomplished, and increase the likelihood of a successful research endeavor.

6.4 What Is/Are Research Question(s) and Why Research Question(s)?

Research questions in a research proposal are specific inquiries or queries that the researcher seeks to answer through the study. They are framed as clear and focused questions that guide the research process and help achieve the research objectives. Research questions serve as the backbone of the research proposal, providing a structured and systematic approach to investigating the research problem.

6.4.1 Characteristics of Research Questions

Clear and Concise: Research questions should be phrased clearly and concisely, avoiding ambiguity or vagueness. They should be specific enough to provide a clear direction for the study.

Relevant and Aligned: Research questions should directly relate to the research problem stated in the proposal. They should align with the research objectives and the overall purpose of the study.

Open-Ended or Specific: Research questions can be open-ended, allowing for exploratory research, or specific, leading to focused investigations. The choice depends on the research design and the level of prior knowledge about the topic.

Feasible and Realistic: Research questions should be feasible and achievable within the constraints of the study, including available resources, time, and access to data.

Guiding the Research Process: Research questions guide the selection of research methods, data collection, and data analysis. They provide a framework for data interpretation and discussion of findings.

6.4.2 Importance of Research Questions

Focus and Direction: Research questions provide a clear focus and direction for the study. They help researchers stay on track and avoid drifting away from the core research objectives.

Structure the Proposal: Research questions form the basis for structuring the research proposal. They often appear in the introduction or literature review to explain the purpose and rationale of the study.

Guide Literature Review: Research questions help in conducting a targeted and relevant literature review. The review revolves around addressing the gaps and knowledge deficits highlighted by the research questions.

Research Design and Methods: The formulation of research questions influences the selection of research design and methods. Whether the study is qualitative, quantitative, or mixed-methods, the research questions shape the choice of appropriate approaches.

Data Collection and Analysis: Research questions determine the type of data to be collected and analyzed. They help in deciding which variables or aspects to focus on during data collection.

Evaluation of Outcomes: Research questions set the criteria for evaluating the research outcomes. They help researchers determine whether the study has successfully answered the questions and achieved its objectives.

Contributions to Knowledge: Research questions drive the creation of new knowledge and insights. By answering the questions, the study may add to the existing literature or challenge prevailing assumptions.

Research questions are essential elements of a research proposal as they provide a structured approach to addressing the research problem. Well-formulated research questions ensure that the study remains focused, feasible, and relevant. They guide the research process from inception to conclusion and lead to meaningful contributions to the field of study.

6.5 Types of Research Question(s)

Research questions in a research proposal can be categorized into two main types: the principal research question and auxiliary research questions.

6.5.1 Principal Research Question

The principal research question is the primary and central question that the study seeks to answer. It represents the main focus of the research and reflects the overarching goal of the study. The principal research question is usually broad and general, encompassing the main topic or theme of the research. The principal research question typically appears at the beginning of the research proposal, in the introduction or background section. It sets the stage for the entire study and guides the formulation of specific objectives and hypotheses. The principal research question is essential in providing clarity about the primary aim of the research and its significance.

Example of a Principal Research Question:

- "What is the impact of social media usage on students' academic performance?"

6.5.2 Auxiliary Research Questions

Auxiliary research questions, also known as sub-questions or secondary questions, are specific and focused inquiries that support and complement the principal research question. These questions break down the main research question into smaller, more manageable components, addressing different aspects or dimensions of the research problem. Auxiliary research questions often stem from the specific objectives of the study. Each objective may be associated with one or more auxiliary research questions. These questions provide a structured and systematic approach to exploring various aspects of the research problem in detail.

Example of Auxiliary Research Questions:

- "What is the frequency of social media usage among students?"
- "How do students perceive the impact of social media on their study habits?"
- "Are there differences in academic performance between students who use social media frequently and those who use it sparingly?"

6.6 Importance of Research Question(s)

The principal research question provides a broad and holistic view of the research focus, while auxiliary research questions help in conducting a more in-depth and comprehensive investigation. Together, they ensure that the research is well-structured, focused, and addresses various dimensions of the research problem. The principal research question guides the overall direction of the study, while auxiliary research questions facilitate the identification of specific data collection methods, analysis techniques, and interpretation of results. The combination of principal and auxiliary research questions helps researchers gain a deeper understanding of the research problem and achieve the research objectives effectively. It is essential to carefully design both types of research questions to ensure that they align with the research problem, research objectives, and theoretical framework. Well-formulated research questions lead to a well-organized and coherent research proposal, making the study more compelling and contributing valuable insights to the field of study.

6.7 How Can We Write Effective Research Question(s)?

Writing effective research questions in a research proposal is critical for formulating a focused and meaningful study. Well-crafted research questions guide the research process, help in data collection and analysis, and provide clear direction for the study. Here are some tips for writing effective research questions in a research proposal:

- Be Clear and Specific: Ensure that each research question is clear, specific, and unambiguous. Avoid vague language or complex sentence structures that may lead to confusion. Each question should address a single aspect of the research problem to maintain focus.
- Relate to the Research Problem: Align the research questions with the main research problem stated in the proposal. Each question should directly contribute to answering the overall research problem and reflect the central theme of the study.
- Use Action-Oriented Language: Frame the research questions using action-oriented language. Use strong verbs that convey the intent of the research, such as "investigate," "examine," "explore," "compare," "identify," "analyze," or "assess."
- Avoid Biased Language: Ensure that the research questions are neutral and unbiased. Avoid leading or suggestive language that could

influence the participants' responses. The questions should be designed to collect objective data.

Reflect on the Research Objectives: The research questions should be closely tied to the specific objectives of the study. Each question should address one or more specific objectives and contribute to achieving the study's overall goals.

Consider Feasibility: Make sure that the research questions are feasible and can be realistically answered within the available resources, time frame, and scope of the study. Avoid overly ambitious questions that may not be achievable.

Mix Different Types of Questions (if applicable): Depending on the research design, mix different types of research questions, such as descriptive, explanatory, comparative, or exploratory questions, to address various aspects of the research problem and provide a comprehensive analysis.

Be Open-Ended or Closed-Ended (if applicable): Depending on the research design and objectives, consider whether the research questions should be open-ended, allowing participants to provide detailed responses, or closed-ended, offering predefined response options.

Align with the Theoretical Framework: Ensure that the research questions are in line with the theoretical framework or existing literature. The questions should address gaps in the current knowledge and contribute to the theoretical understanding of the research problem.

Seek Feedback and Refine: Share the research questions with colleagues, mentors, or advisors to get feedback. Revise and refine the questions based on their input to ensure clarity, relevance, and coherence.

Writing effective research questions is an iterative process that requires careful consideration and attention to detail. Strong research questions contribute to a well-structured and impactful research proposal, demonstrating the researcher's ability to address the research problem with precision and thoroughness.

CHAPTER 7

Rationale of the Study of a Research Proposal

ABSTRACT
This chapter delves into the profound significance of the "rationale of the study" within a research proposal. It begins by elucidating the essence of this section, which provides the compelling reasoning behind initiating the research endeavor. The multifaceted objectives of the "rationale of the study" are explored, highlighting its pivotal role in establishing the study's necessity. In addition, the chapter dissects the justification for the research, encompassing its relevance to existing gaps and issues. It emphasizes the interplay between research objectives, hypotheses, and rationale, demonstrating their symbiotic relationship. The integration of a comprehensive literature review is showcased as a vital component, illustrating the proposal's alignment with existing scholarship. Additionally, the practical feasibility of the study is examined, acknowledging the importance of viability in real-world implementation. The chapter extends its focus to the potential contributions of the research to knowledge and societal impact, as well as its relevance to policy initiatives. Through a holistic exploration, the chapter guides researchers in constructing a compelling rationale that substantiates the study's significance.

KEYWORDS: Rationale of the study, research objectives, hypotheses, literature review, feasibility, societal impact, policy link

7.1 What Is the Meaning of the Rationale of the Study?

The rationale of the study in a research proposal is a critical section that provides a clear and convincing explanation for why the research is worth undertaking. It serves as the justification and basis for the research, outlining the reasons, significance, and potential contributions of the proposed study. The rationale establishes the relevance and importance of the research, convincing the readers, such as supervisors, funding agencies, or research committees, that the research is necessary and will address a gap or problem in the existing body of knowledge.

7.2 Objectives of the Rationale of the Study

7.2.1 Identifying the Research Gap

The research gap is the lack of knowledge, information, or understanding in a specific area of research that needs to be addressed or explored further. Identifying the research gap is a crucial first step in justifying the need for the proposed study. Here's a more detailed explanation of this point:

> The Context of Research: The rationale starts by providing the context of the research topic within the existing literature and field of study. It outlines the current state of knowledge, theories, and debates related to the research area. This context highlights the importance of the topic and sets the stage for identifying the research gap.
>
> Reviewing Previous Studies: To identify the research gap, researchers conduct a comprehensive literature review. This involves reviewing relevant scholarly articles, books, reports, and other sources to understand the existing body of knowledge on the topic. The literature review helps researchers become familiar with the key concepts, findings, and research questions that have been explored in previous studies.
>
> Critical Evaluation: Researchers critically evaluate the existing literature to pinpoint limitations, unanswered questions, or areas that require further investigation. This evaluation may reveal inconsistencies in findings, methodological flaws, or conflicting theories that indicate a research gap. Additionally, researchers may identify recent developments or emerging trends that have not been adequately addressed in previous research.
>
> Conceptual and Theoretical Gaps: The research gap can be conceptual, theoretical, or empirical. Conceptual gaps refer to areas where the literature lacks a clear definition, conceptual framework, or theoretical basis for understanding the research topic. Theoretical gaps arise when existing theories fail to provide a comprehensive explanation for certain phenomena or fail to address new developments in the field.

- Empirical Gaps: Empirical gaps are related to the lack of empirical evidence or data on specific aspects of the research topic. For example, existing studies may have focused on certain populations, regions, or periods, leaving other groups or contexts underexplored. Empirical gaps can also occur when existing research has used limited methodologies or data collection techniques, leaving room for further investigation using different approaches.
- Unanswered Research Questions: The research gap may be characterized by unanswered research questions that have not been adequately addressed in previous studies. These questions may arise from conflicting findings, inconclusive results, or discoveries that require further investigation.
- Importance of Filling the Gap: The rationale explains the significance of addressing the research gap. Researchers articulate why closing this gap is crucial for advancing knowledge, theory, or practice in the field. They may discuss the potential implications of filling the gap, such as contributing to policy development, improving interventions, or enhancing understanding in a specific area.

By clearly identifying the research gap in the rationale, researchers demonstrate their familiarity with the existing literature, their ability to critically analyze previous studies, and their motivation to contribute meaningfully to the field. A well-justified research gap strengthens the rationale of the study and lays the foundation for formulating specific research objectives and hypotheses in the research proposal.

7.2.2 Significance and Relevance

After identifying the research gap, it is crucial to explain why addressing this gap is important and how the proposed study will contribute to the field of knowledge. Demonstrating the significance and relevance of the research enhances the justification for undertaking the study and highlights its potential impact. Here's a more detailed explanation of this point:

- Addressing Knowledge Gaps: The significance of the research lies in its potential to address the identified research gap and contribute to the existing body of knowledge. By filling this gap, the study aims to provide new insights, extend theoretical frameworks, or validate and refine existing theories. Researchers should explain how their study builds upon and complements previous research, leading to a deeper understanding of the research topic.
- Practical Applications: The relevance of the research lies in its potential practical applications and implications. Researchers should articulate how the study's findings can be applied in real-world settings to solve problems, improve practices, or inform decision-making. Practical relevance is particularly important in applied research fields where the ultimate goal is to make a positive impact on society or specific industries.
- Contributions to Policy and Practice: If the research has policy implications, researchers should highlight how the study's outcomes can inform policy development, implementation, or evaluation. Policymakers often rely on evidence-based research to make informed decisions, and a well-justified research proposal can strengthen its potential to influence policy direction.
- Addressing Societal Challenges: Research that addresses pressing societal challenges or global issues holds significant relevance. By demonstrating how the study's outcomes can contribute to tackling societal problems, researchers can attract support and funding from stakeholders and funding agencies interested in promoting social welfare and positive change.
- Advancing Methodological Approaches: The significance of the research may also lie in its methodological contributions. If the study proposes innovative or novel research methods, data collection techniques, or analytical approaches, researchers should highlight how these advancements can benefit the wider research community and improve the quality of future studies.
- Building a Knowledge Network: Research that contributes to building a comprehensive knowledge network in a specific field

or subfield is considered highly relevant. By connecting diverse findings and synthesizing research from various sources, the study enriches the collective knowledge base and fosters collaboration among researchers.

Addressing Global Challenges: Research that addresses global challenges, such as climate change, public health crises, or technological advancements, is particularly significant. These studies have the potential to impact multiple regions, countries, or even the entire world, making them highly relevant to global audiences.

Catalyzing Further Research: The proposed study's significance may also lie in its potential to inspire and catalyze further research on related topics. By paving the way for future investigations, the research becomes a building block for a broader research agenda and contributes to the growth of the field.

Impact on Stakeholders: Researchers should consider the interests and needs of various stakeholders, such as communities, organizations, or industries, and explain how the study's findings can benefit these groups. Demonstrating the impact of the research on stakeholders can foster support and collaboration.

Articulating the significance and relevance of the research in the rationale of the proposal not only strengthens its justification but also underscores the researcher's passion and commitment to addressing meaningful research questions. By showcasing the potential contributions and impacts of the study, researchers increase the likelihood of obtaining support, funding, and ethical approval for their research project.

7.3 Justification for the Study

The justification section in the rationale of a research proposal outlines the reasons behind the research and its potential contributions to the field. It aims to convince readers, such as supervisors, funding agencies, or research committees, that the research is significant and will address a relevant problem or gap in knowledge. Here's a more detailed explanation of this point:

> Research Context and Background: The justification begins by providing a clear and concise overview of the research context and background. It introduces the research topic, its relevance to the field of study, and any existing issues or gaps in knowledge. This context sets the stage for explaining the need for the proposed study.
>
> Identifying Research Questions and Objectives: The research proposal's justification should explicitly state the research questions and objectives that the study aims to address. These research questions should directly align with the identified research gap and should be framed in a way that emphasizes their significance and potential impact.
>
> Addressing Societal Needs or Problems: Justifying the research involves highlighting how the study addresses societal needs or real-world problems. Researchers should emphasize how the proposed research has practical applications and can make a positive impact on the lives of individuals, communities, or industries.
>
> Advancing Knowledge and Theory: Researchers should demonstrate how the proposed study contributes to the advancement of knowledge and theory in the field. This may involve building upon existing theories, testing hypotheses, or proposing new conceptual frameworks to explain phenomena.
>
> Potential Contributions to Practice: In addition to advancing theoretical knowledge, the justification should discuss how the study's findings can be applied to practical settings. Researchers can explore how their research results can inform decision-making, improve practices, or lead to innovative solutions.
>
> Filling Research Gaps: Justifying the study involves explicitly explaining how the proposed research addresses specific gaps in the existing literature. Researchers should discuss why previous studies have not adequately addressed the research questions and how their study provides a fresh perspective or innovative approach.

- Addressing Current Relevance: The research proposal's justification should highlight the current relevance of the research topic. This may involve linking the study to ongoing debates, contemporary issues, or recent developments in the field.
- Building on Preliminary Work or Pilot Studies: If applicable, researchers can discuss any preliminary work, pilot studies, or exploratory research they have conducted related to the proposed study. Building on prior work strengthens the rationale and demonstrates a thoughtful and informed approach to the research.
- Potential for High Impact: Researchers should articulate the potential impact of the study's findings on the field and beyond. Emphasizing the significance of the research results can attract support and interest from stakeholders, funding agencies, and the broader research community.
- Alignment with Organizational or Institutional Priorities: For research proposals seeking funding from specific institutions or organizations, it is essential to demonstrate how the study aligns with their priorities and objectives. Researchers should explain how their research can contribute to the mission and goals of the funding agency.

In summary, justifying the research proposal is about making a compelling case for why the proposed study is necessary, relevant, and valuable. By clearly presenting the research questions, objectives, and potential contributions, researchers can enhance the persuasiveness of their proposal and increase the likelihood of obtaining support and approval for their research project.

7.4 Research Objectives and Hypotheses

Research objectives and hypotheses are crucial components of a well-structured research proposal, as they outline the specific aims and expectations of the study. Clearly defining these elements helps in designing the research and measuring its success. Here's a more detailed explanation of this point:

Defining Research Objectives: Research objectives are specific statements that outline the intended outcomes and goals of the study. These objectives provide a clear direction for the research and guide the entire research process, from data collection to analysis and interpretation of results. Well-defined research objectives are essential for maintaining focus and ensuring that the study remains aligned with its purpose.

Types of Research Objectives: Research objectives can be broadly categorized into general and specific objectives. General objectives outline the overall goal of the study, while specific objectives break down the general goal into smaller, measurable steps. Specific objectives are often more detailed and provide a roadmap for achieving the general objective.

Linking Objectives to the Rationale: The research proposal should establish a clear link between the research objectives and the rationale. Each research objective should directly address the research gap and the significance of the study. Demonstrating this alignment strengthens the rationale and shows how the proposed study directly contributes to filling the identified gap in knowledge.

Formulating Hypotheses: Hypotheses are testable statements that predict the relationship between variables or the expected outcomes of the research. In quantitative research, hypotheses are typically formulated as null (H_o) and alternative (H_a) hypotheses. Null hypotheses propose no significant relationship between variables, while alternative hypotheses suggest that there is a significant relationship.

Role of Hypotheses in Quantitative Research: In quantitative research, hypotheses play a critical role in guiding data analysis and statistical testing. They allow researchers to make objective predictions about the results and evaluate whether the data supports or refutes the proposed relationships between variables. Well-defined

and testable hypotheses enhance the rigor of quantitative research and facilitate the interpretation of findings.
- Research Questions and Qualitative Research: In qualitative research, research questions often take the place of hypotheses. Research questions are open-ended inquiries that guide the investigation and exploration of a research topic. Unlike hypotheses, research questions do not propose specific relationships between variables but rather aim to understand complex phenomena from the perspective of participants.
- Clarity and Specificity: Both research objectives and hypotheses should be formulated with clarity and specificity. Vague or ambiguous statements may lead to confusion during the research process and hinder the ability to draw meaningful conclusions.
- Feasibility and Realism: Researchers should ensure that their research objectives and hypotheses are feasible and realistic within the scope of the study. Unrealistic or overly ambitious objectives may lead to difficulties in data collection and analysis.
- Alignment with Research Design and Methodology: The research objectives and hypotheses should align with the chosen research design and methodology. The data collected and analyzed should be directly related to the research objectives to ensure that the study addresses the intended research questions.

By formulating clear and well-defined research objectives and hypotheses, researchers provide a roadmap for their study, set specific targets to achieve, and facilitate the evaluation of the study's success. These elements demonstrate the researcher's foresight and strategic thinking, enhancing the overall quality and credibility of the research proposal.

7.5 Literature Review

The literature review is a critical component that evaluates existing research and knowledge related to the proposed study. It serves to situate the research within the context of previous work, identify gaps in knowledge, and establish the need for the proposed study. Here's a more detailed explanation of this point:

> Purpose of the Literature Review: The literature review in a research proposal serves multiple purposes. It demonstrates the researcher's familiarity with the existing body of knowledge on the research topic and shows that the proposed study builds upon previous work. The literature review also highlights the gaps, limitations, and controversies in the literature, which justify the need for the proposed research.
>
> Scope and Depth of the Literature Review: A comprehensive literature review covers relevant scholarly articles, books, conference proceedings, and other reputable sources related to the research topic. The depth of the review will depend on the complexity of the research area and the availability of published literature. A good literature review critically analyzes and synthesizes previous studies to present a coherent and up-to-date understanding of the research topic.
>
> Organizing the Literature Review: The literature review should be organized thematically or chronologically, depending on the research question and the relationship between the reviewed studies. A thematic review groups studies based on common themes or theoretical frameworks, while a chronological review presents the historical development of the research topic.
>
> Identifying Key Concepts and Theories: The literature review should identify the key concepts, theories, and models that are relevant to the research topic. By presenting a clear theoretical framework, the rationale of the study establishes its theoretical foundation and aligns with established scholarly debates.
>
> Identifying Gaps in Knowledge: A primary objective of the literature review is to identify gaps in the existing knowledge base. These gaps could be areas where conflicting findings exist, questions that have not been adequately answered, or new developments that require

further exploration. The identification of research gaps is essential for justifying the need for the proposed study.

Building a Logical Argument: The literature review should build a logical argument that connects the existing research with the proposed study. Researchers should demonstrate how the proposed research fits into the broader research landscape and complements or extends the existing literature.

Critically Evaluating Previous Studies: A well-structured literature review critically evaluates the strengths and weaknesses of previous studies. Researchers should assess the methodology, sampling, data collection, and analysis techniques used in previous research to identify any limitations or biases.

Highlighting Supporting Evidence: The literature review should highlight studies that support the research objectives and hypotheses proposed in the rationale. Demonstrating that previous research findings align with the anticipated outcomes of the study strengthens the justification for the research.

Establishing Credibility and Expertise: By conducting a thorough literature review, researchers demonstrate their credibility and expertise in the field. This reassures readers that the proposed study is well-informed and grounded in existing scholarship.

Synthesizing Diverse Perspectives: The literature review should synthesize diverse perspectives and conflicting findings to provide a balanced overview of the research topic. This can lead to a more nuanced understanding of the complexities surrounding the research area.

Potential for Innovation: In addition to identifying gaps, the literature review may also reveal opportunities for innovation and novel contributions. Researchers should discuss how their proposed study addresses unique aspects of the research topic that have not been explored previously.

A well-written literature review strengthens the rationale of a research proposal by demonstrating the research's relevance, originality, and theoretical grounding. It provides the necessary context for the proposed study and convinces readers of the importance of undertaking the research. A comprehensive literature review is an essential part of building a persuasive case for the research proposal and ensures that the study aligns with current knowledge and addresses significant gaps in the field.

7.6 Practical Feasibility

Practical feasibility refers to the extent to which the proposed research can be realistically and successfully carried out within the available resources, constraints, and timelines. Demonstrating practical feasibility is essential to convince stakeholders, funders, and reviewers that the proposed study is viable and achievable. Here's a more detailed explanation of this point:

> Resource Availability: Practical feasibility begins by assessing the availability of necessary resources for the research. These resources include financial support, personnel, research facilities, access to data, equipment, and any other materials required for the study. Researchers should demonstrate that the necessary resources are either already secured or reasonably obtainable within the planned timeframe.
>
> Time Constraints: The rationale should address the time constraints associated with the research project. Researchers need to show that the proposed study can be completed within a realistic timeframe and that the timeline is aligned with the research objectives. A well-structured research plan with clear milestones helps demonstrate time management and feasibility.
>
> Research Team Expertise: Practical feasibility involves evaluating the expertise and skills of the research team. Researchers should highlight their qualifications, experience, and capability to conduct the proposed study. If the study requires specific technical or

methodological expertise, the rationale should demonstrate that the research team possesses the necessary competencies or has access to collaborators with the required skills.

Research Design and Methodology: The chosen research design and methodology should be appropriate and feasible for addressing the research questions or objectives. Researchers should explain how the selected approach aligns with the available resources and the scope of the study. Justifying the research design demonstrates careful planning and thoughtful consideration of how to collect and analyze data effectively.

Sampling and Recruitment: Practical feasibility involves addressing how participants or subjects will be recruited and how the sample size will be determined. Researchers should discuss potential challenges related to sampling and recruitment and provide strategies to mitigate them. Adequate representation and access to the target population are essential considerations for practical feasibility.

Data Collection Methods: The rationale should justify the selected data collection methods and tools, considering the practical aspects of data collection. Researchers should demonstrate that the chosen methods are suitable for capturing the required data and can be feasibly implemented within the research context.

Data Analysis Techniques: Similarly, researchers need to justify the data analysis techniques they plan to use. The rationale should show that the selected analysis methods are compatible with the type of data collected and that the necessary resources, such as statistical software or expertise, are available.

Ethical Considerations: Practical feasibility also involves addressing ethical considerations related to the research. Researchers should demonstrate that they have considered and will adhere to ethical guidelines and obtain necessary ethical approvals for the study.

Potential Challenges and Contingency Plans: In assessing practical feasibility, researchers should identify potential challenges and risks that may arise during the research process. The rationale should discuss how these challenges will be addressed and provide contingency plans to deal with unforeseen circumstances.

By thoroughly addressing the practical feasibility of the research, the rationale provides evidence of the researcher's careful planning, resource management, and foresight. Demonstrating practical feasibility enhances the credibility of the research proposal and assures stakeholders that the proposed study can be executed successfully within the available resources and timelines. A feasible research plan increases the likelihood of obtaining support, funding, and ethical approval for the research project.

7.7 Contributions to Knowledge

Articulating the potential contributions to knowledge is a crucial aspect of the rationale, as it highlights the unique and original aspects of the proposed study. Demonstrating how the research will advance existing knowledge and fill gaps in the literature is essential to justify the significance and relevance of the study. Here's a more detailed explanation of this point:

- Identifying Research Gap: The rationale of the research proposal should start by identifying the specific research gap or problem in the existing literature. Researchers should explain why the identified gap is significant and why it requires further investigation. This gap serves as the foundation for the potential contributions to knowledge that the proposed study aims to make.
- Novelty and Originality: Researchers need to highlight the novelty and originality of their proposed study. This involves explaining how the research goes beyond existing studies and offers new insights, perspectives, or methodologies. Demonstrating the originality of the study enhances its value and distinguishes it from previous work.
- Addressing Controversies or Contradictions: If the proposed study aims to address controversies or contradictions in the literature,

the rationale should clearly state how it plans to do so. Resolving these discrepancies can lead to a more comprehensive and refined understanding of the research topic.

Extending Theoretical Frameworks: The rationale should discuss how the proposed study extends existing theoretical frameworks or models. Researchers can explain how their research builds upon established theories and adds new dimensions to them. Extending theoretical frameworks contributes to the evolution of the field and can lead to broader applications of existing concepts.

Replicating and Validating Findings: In some cases, the rationale may emphasize the importance of replicating and validating findings from previous studies. Replication studies are valuable contributions to knowledge as they verify the robustness and reliability of existing findings.

Developing Practical Applications: The rationale should discuss how the proposed research can have practical applications or implications. Researchers can explain how their findings can be applied in real-world contexts to address specific challenges or improve practices in relevant domains.

Creating New Data or Datasets: If the research involves the creation of new data or datasets, the rationale should emphasize the value of such resources. New data can serve as valuable assets for other researchers in the field, promoting future investigations and advancing knowledge.

Promoting Interdisciplinary Connections: The proposed study may have the potential to bridge gaps between different disciplines or fields. Researchers can explain how their work fosters interdisciplinary connections and contributes to a more holistic understanding of the research topic.

Enabling Future Research: By providing valuable insights and paving the way for further investigations, the proposed study can stimulate future research and open up new avenues of inquiry. Researchers should discuss how their work lays the groundwork for future studies and research agendas.

Potential Policy Implications: If the research has policy implications, the rationale should highlight how the study's findings can inform decision-making and policy development. Policymakers often rely on research evidence to make informed choices, and a research proposal with clear policy implications can attract support and funding.

By emphasizing the contributions to knowledge in the rationale, researchers showcase the intellectual significance of their proposed study. Demonstrating how the research will advance existing knowledge, fill gaps, and bring new perspectives to the field enhances the overall quality and importance of the research proposal. It also conveys the researcher's awareness of the broader implications of the study and its potential to make a meaningful impact on the academic or practical landscape.

7.8 Societal Impact

Societal impact refers to the potential effects and benefits that the proposed study may have on individuals, communities, organizations, or society as a whole. Demonstrating the societal impact is essential to justify the relevance and significance of the research and to show that it goes beyond academic contributions. Here's a more detailed explanation of this point:

Addressing Real-World Problems: The rationale of the research proposal should highlight how the proposed study addresses real-world problems or challenges. Researchers should explain how the findings of the study can contribute to solutions for practical issues faced by individuals, communities, or organizations. Addressing such problems enhances the societal relevance of the research.

Improving Quality of Life: Researchers should discuss how the research outcomes can potentially improve the quality of life for individuals or groups affected by the research topic. This could involve advancements in healthcare, social services, education, technology, or other domains that directly impact people's well-being.

Informing Policy and Decision-Making: If the research has implications for policy development or decision-making, the rationale should highlight this potential impact. Policymakers often rely on evidence-based research to make informed choices, and a research proposal with clear policy implications can attract support and funding.

Contributing to Sustainable Development: The research proposal should demonstrate how the study aligns with the principles of sustainable development. This includes considering the environmental, social, and economic impacts of the research and ensuring that it contributes positively to sustainable practices.

Empowering Marginalized or Vulnerable Groups: Researchers should discuss how their study can empower marginalized or vulnerable groups by providing insights or solutions to their specific challenges. This may involve addressing issues related to social justice, equity, or human rights.

Cultural and Historical Preservation: In some cases, the proposed research may focus on cultural or historical preservation. Researchers can explain how their work contributes to the documentation and protection of cultural heritage or historical knowledge.

Fostering Innovation and Creativity: The rationale should discuss how the proposed study can foster innovation and creativity in relevant fields. Research that explores new ideas, methods, or technologies can inspire further advancements and contribute to the progress of society.

Public Engagement and Awareness: Researchers can discuss how they plan to engage with the public and raise awareness about the research topic. Public engagement can promote a better understanding of the research and its potential impact on society.

Educational and Training Opportunities: The research proposal can highlight how the study may create educational or training opportunities for students, professionals, or community members. This could involve skill development, capacity building, or knowledge transfer.

Promoting Inclusivity and Diversity: If the research is focused on issues related to diversity and inclusion, the rationale should explain how the study contributes to promoting inclusivity and understanding among different groups or communities.

By emphasizing the societal impact of the proposed research, researchers demonstrate their commitment to conducting research that goes beyond academic boundaries and contributes to the betterment of society. A research proposal that addresses societal concerns and has potential implications for real-world issues is more likely to gain support from stakeholders, funders, and review committees, as it demonstrates the study's relevance and significance to a broader audience.

8.9 Policy Link

The policy link refers to the connection between the proposed research and the development or modification of policies or guidelines. Demonstrating a clear policy link is crucial as it showcases the potential impact of the research on policymaking and its relevance to addressing societal challenges. Here's a more detailed explanation of this point:

Identifying Relevant Policies: The rationale of the research proposal should identify existing policies or guidelines that are directly related to the research topic. Researchers should explain how their proposed study aligns with these policies and contributes to addressing the goals and objectives outlined in them.

Filling Policy Gaps: In some cases, the research proposal may identify gaps or limitations in existing policies or guidelines. Researchers can explain how their study aims to fill these gaps by providing evidence-based recommendations or insights to inform policy development or revision.

- Supporting Evidence-Based Policymaking: The rationale should emphasize the potential of the proposed research to support evidence-based policymaking. Policymakers often rely on research evidence to make informed decisions, and a well-structured research proposal can serve as a valuable resource for guiding policy choices.
- Engaging Stakeholders and Policymakers: Researchers can discuss how they plan to engage with stakeholders and policymakers throughout the research process. Involving relevant stakeholders can enhance the policy relevance of the study and ensure that the research findings are applicable and practical for policymaking.
- Policy Implications of Findings: The rationale should outline the policy implications of the anticipated research findings. Researchers should explain how the study's outcomes can be translated into actionable policy recommendations and strategies.
- Promoting Evidence-Driven Reforms: The proposed research can highlight the need for evidence-driven reforms in specific areas. By providing robust evidence and data, the study can advocate for policy changes that lead to improved practices and outcomes.
- Informing Public Debates: The rationale can discuss how the research findings can contribute to public debates on relevant policy issues. By disseminating research results to the public, researchers can raise awareness and foster informed discussions on matters of public interest.
- Advocating for Change: If the research is focused on advocacy or social change, the rationale should explain how the study aims to influence policy decisions and promote positive transformations in society.
- Policy Recommendations: The research proposal can include preliminary policy recommendations based on existing knowledge and the research gap identified. These recommendations can demonstrate the potential practical applications of the proposed study.
- Measuring Policy Impact: In some cases, the rationale may discuss how the research aims to measure the impact of existing policies or

interventions. Understanding the effectiveness of policies can lead to evidence-based improvements and better outcomes.

By establishing a strong policy link in the rationale, researchers highlight the potential of their research to make a tangible difference in society. Policymakers and funding agencies often prioritize studies that offer relevant and applicable insights to address societal challenges and inform policymaking. A research proposal with a clear policy link demonstrates that the study goes beyond theoretical exploration and has the potential to shape practical solutions and policy reforms.

7.10 Rationale of the Study of a Research Proposal

Topic: "Exploring the Impact of Community Gardens on Urban Well-being in Durban City in South Africa"
In the context of Durban City in South Africa, this study seeks to address the repercussions of rapid urbanization and changing lifestyles that have severed the connection between individuals and nature. This disconnect has contributed to escalating stress levels and a decline in overall well-being. The rationale for the study lies in recognizing the potential of community gardens as a remedy to this issue. The investigation is deemed essential for fostering healthier urban environments and understanding how participation in community gardening can positively influence individual well-being. The hypotheses put forward suggest that active engagement in community gardening will correlate positively with indicators of mental health, including reduced stress levels, an increased sense of community, and enhanced overall life satisfaction.

The practical feasibility of the study is assessed by considering the availability of suitable spaces for community gardens, community members' willingness to participate, and the logistical aspects of establishing and maintaining these gardens. The section outlines strategies to address potential challenges, highlighting the scalability of community gardening initiatives. The study aims to contribute to existing knowledge by providing empirical evidence on the relationship between community gardening and well-being, intending to fill gaps in the current literature by exploring the nuanced ways in which community engagement with green spaces positively influences the mental and social aspects of urban life.

With a focus on societal impact, the study envisions the creation of healthier and more resilient communities. It anticipates that the findings will inform urban planning strategies, encouraging the integration of community gardens into city landscapes to enhance the overall well-being of residents. The research establishes a direct link to policy considerations by advocating for urban policies that prioritize the creation and maintenance of community gardens, proposing that the integration of community gardens should be a crucial element in urban planning frameworks to promote community health and well-being in Durban City.

In conclusion, this chapter has provided a comprehensive and step-by-step methodological description for crafting a research proposal. Throughout the chapter, we have explored various critical elements that are essential for constructing a well-structured and compelling research proposal. The chapter began by discussing the importance of understanding the research problem and formulating clear research objectives and questions. We delved into the significance of conducting a thorough literature review to establish the conceptual framework and identify gaps in existing knowledge. By doing so, we ensure that the proposed study adds value to the academic landscape and addresses relevant research questions. Furthermore, the chapter emphasized the significance of choosing appropriate research methods and data collection tools that align with the research objectives. Whether quantitative, qualitative, or mixed methods, the selected approach should be carefully justified based on the research design and feasibility.

CHAPTER 8

Step-by-Step Methodology of a Research Proposal

ABSTRACT
This chapter presents a comprehensive roadmap for formulating the methodology section of a research proposal. It begins by emphasizing the pivotal aspect of locating the study within a specific context, which sets the stage for subsequent methodological choices. Different research approaches are scrutinized, providing insights into aligning the study's design with its objectives. Furthermore, the chapter navigates the diverse range of research methods available, assisting researchers in selecting the most suitable techniques for their study. Exploration into data collection tools and instrument development guides researchers in creating effective means to gather data. The significance of sampling strategies and selecting respondents is discussed to ensure robust data collection. Moreover, data measurement techniques are examined, followed by a comprehensive overview of data analysis methods. The chapter underscores the importance of data validity and reliability, highlighting these as key elements in ensuring research credibility. Ethical considerations and potential limitations are addressed, fostering a well-rounded methodology that accounts for both the study's scope and its ethical implications.

KEYWORDS: Methodology, research approach, research methods, data collection, sampling, data analysis, validity, reliability, ethical issues

The methodology section of a research proposal outlines the systematic approach and procedures that will be used to conduct the study. It describes the location of the study, the research approach (quantitative, qualitative, or mixed), the research methods (experimental, survey, case study, ethnographic, or content analysis), data collection tools (interviews, observations, in-depth case interviews, focus group discussions, key informant interviews, participatory rural appraisal, discourse analysis), data collection instrument development (schedule, guideline, or checklist), sampling and respondents, data measurement, data analysis techniques, ethical issues, and the limitations of the study. Each component is crucial in ensuring the research is well-designed, rigorous, and ethically conducted.

8.1 Location of the Study

In the methodology section of a research proposal, providing a detailed and comprehensive description of the location of the study is crucial for contextualizing the research and understanding the environment in which the data will be collected. The location of the study refers to the geographical area or setting where the research will be conducted. This section helps readers grasp the significance of the chosen location and how it relates to the research objectives.

Importance of Describing the Location:

- Relevance and Context: Explaining the location of the study helps readers understand why a particular setting was selected. It provides context and relevance to the research by highlighting the connection between the research problem and the location.
- Generalizability: The location can influence the generalizability of the findings. Describing the location allows readers to assess how applicable the research results might be to other similar settings or populations.
- Spatial Considerations: The geographic location may have spatial implications for the research, especially in studies that involve geographical differences, climate variations, or regional disparities.
- Accessibility and Resources: Describing the location helps evaluate the accessibility of the study area and the availability of necessary resources (e.g. data, participants, and facilities) for conducting the research.
- Cultural and Social Context: The location may have unique cultural, social, or political characteristics that can impact the research. Understanding the cultural context is essential in designing appropriate data collection tools and addressing potential biases.
- Logistical Challenges: Some locations may present logistical challenges (e.g. remote areas and conflict zones), and it is essential to address how these challenges will be managed during the research.

- Previous Studies: Mentioning any prior research conducted in the same or similar locations provides a foundation for the current study and highlights any gaps that the research aims to address.

How to Describe the Location:

- Geographical Details: Include specific geographical details, such as the country, region, city, or community, where the study will take place. Provide a brief overview of the location's demographics and key characteristics.
- Justification: Explain why the chosen location is ideal for addressing the research problem. Mention any unique aspects of the location that make it relevant to the study.
- Setting and Context: Describe the setting where data collection will occur. For example, if it is an educational institution, a healthcare facility, a rural village, or an online platform.
- Accessibility and Permissions: Address the accessibility of the location and how you obtained the necessary permissions or approvals to conduct the research.
- Temporal Aspect: Mention any temporal considerations related to the location, such as seasonal variations, historical significance, or specific events that may impact the research.
- Geospatial Tools (if applicable): If the research involves geospatial data or mapping, mention the tools or technologies used to analyze spatial patterns or trends.
- Scope and Boundaries: Clearly define the boundaries of the study area and explain why these boundaries were chosen. This is especially important for studies that cover extensive regions.

By providing a comprehensive description of the location, researchers can demonstrate the rationale behind their choice of study area and build a strong foundation for the methodology of the research proposal. This information helps readers understand the context in which the research will take place and contributes to the overall credibility and validity of the study.

8.2 Research Approach

Specify the research approach that will be used in the study, such as quantitative, qualitative, or mixed methods. Justify the choice of approach based on the research objectives and the nature of the research problem. Explain how the selected approach will enable you to gather the necessary data and address the research questions. The research approach is a fundamental aspect of the methodology in a research proposal, as it outlines the overall strategy and perspective that will guide the study's design, data collection, and analysis. There are three main research approaches: quantitative, qualitative, and mixed methods. Each approach offers unique strengths and advantages, and the choice depends on the research objectives, the nature of the research problem, and the type of data needed to answer the research questions.

8.2.1 Quantitative Research Approach

The quantitative research approach involves the systematic collection and analysis of numerical data to quantify phenomena and establish patterns or relationships. It focuses on generating statistically valid and generalizable findings. This approach is commonly used in studies that aim to measure variables, test hypotheses, and make predictions.
Key Characteristics:

- Objective Measurement: Quantitative research relies on standardized instruments and measures to ensure objectivity and precision in data collection.
- Large Sample Size: Quantitative studies often require larger sample sizes to achieve statistical power and enhance the representativeness of the results.
- Statistical Analysis: Data analysis in quantitative research involves various statistical techniques, such as inferential statistics, correlations, and regression analysis.

Generalizability: The findings from quantitative research are typically more generalizable to larger populations due to the use of random sampling and statistical procedures.

8.2.2 Qualitative Research Approach

The qualitative research approach aims to understand and interpret social phenomena in their natural context. It involves the collection of non-numerical data, such as textual or visual information, to gain insights into the meanings, experiences, and perspectives of participants. Qualitative research is particularly suited for exploring complex, nuanced, and context-bound aspects of a research problem.
Key Characteristics:

- In-depth Exploration: Qualitative research emphasizes in-depth exploration of individuals' thoughts, feelings, behaviors, and social interactions.
- Small Sample Size: Qualitative studies typically involve smaller sample sizes, as the focus is on the richness and depth of data rather than statistical representativeness.
- Data Interpretation: Data analysis in qualitative research involves coding, categorizing, and identifying themes or patterns to develop a comprehensive understanding of the research problem.
- Contextual Understanding: Qualitative research provides a deeper understanding of the social and cultural context in which the phenomena occur.

8.2.3 Mixed Methods Research Approach

The mixed methods research approach combines both quantitative and qualitative methods within a single study. This approach aims to capitalize on the strengths of both methods and provides a more comprehensive understanding of the research problem.

Key Characteristics:

- Triangulation: Mixed methods research uses multiple data sources to triangulate findings and enhance the credibility and validity of the results.
- Complementarity: Quantitative and qualitative data are used to complement each other, offering a more holistic view of the research problem.
- Sequential or Concurrent Design: Mixed methods studies can be conducted sequentially (i.e. one phase follows the other) or concurrently (i.e. both phases are conducted simultaneously).

8.2.4 Choosing the Research Approach

The choice of research approach is influenced by various factors, including the research objectives, research questions, availability of resources, and the researcher's philosophical stance. Researchers should carefully consider the strengths and limitations of each approach and select the one that best aligns with the nature of the research problem and the desired outcomes.

By clearly articulating the research approach in the research proposal, the researcher sets the foundation for the study's methodology. The chosen approach shapes the data collection methods, analysis techniques, and interpretation of results, ensuring that the study is well-designed and capable of producing valuable insights. It also allows reviewers or readers of the proposal to understand the researcher's methodological choices and evaluate the suitability of the research approach for the stated research objectives.

8.3 Research Methods

Detail the specific research methods that will be employed, such as experimental design, surveys, case studies, ethnographic observations, or content analysis. Provide a rationale for selecting these methods, and describe how they align with the research approach. Explain how each method will contribute to achieving the research objectives. The research methods section of a research proposal outlines the specific techniques and procedures that will be employed to collect and analyze data. It involves selecting appropriate research methods that align with the chosen research approach (quantitative, qualitative, or mixed methods) and addressing the research questions effectively. Different research methods offer unique advantages and are suitable for different types of research inquiries. Here are further insights into the various research methods commonly used in research proposals:

8.3.1 Experimental Research Method

Experimental research is a quantitative method that involves the manipulation of one or more independent variables to observe their effect on a dependent variable while controlling for confounding factors. It is commonly used to establish cause-and-effect relationships. Experimental studies use random assignment of participants to experimental and control groups.
Key Characteristics:

- Controlled Environment: Experiments are conducted in controlled environments, allowing researchers to minimize external influences and isolate the effects of the independent variable(s).
- Randomization: Random assignment helps ensure that participants in the experimental and control groups are equivalent at the beginning of the study, reducing bias.
- Quantitative Data: Data collected in experimental research is typically numerical and amenable to statistical analysis.

8.3.2 Survey Research Method

Surveys are quantitative research methods that involve collecting data through standardized questionnaires or structured interviews. Surveys aim to gather information from a large number of respondents to generalize findings to a broader population.

Key Characteristics:

- Questionnaires or Interviews: Surveys use questionnaires or structured interviews to gather data from participants.
- Large Sample Size: Surveys often require a large sample size to achieve statistically significant results.
- Quantitative Data: Survey data is usually numerical and can be analyzed using statistical methods.

8.3.3 Case Study Research Method

Case studies are qualitative research methods that focus on an in-depth examination of a single individual, group, organization, or phenomenon. Case studies provide detailed and context-rich insights into complex social phenomena.

Key Characteristics:

- In-depth Exploration: Case studies involve extensive data collection through interviews, observations, and document analysis to provide a comprehensive understanding of the case.
- Holistic Perspective: Researchers analyze the case as a whole, considering multiple variables and their interplay.
- Contextual Understanding: Case studies emphasize the importance of understanding the unique context and intricacies of the case.

8.3.4 Ethnographic Research Method

Ethnography is a qualitative research method used to study cultures and communities. Researchers immerse themselves in the culture or community being studied to gain a deep understanding of the participants' perspectives, behaviors, and experiences.

Key Characteristics:

- Participant Observation: Ethnographic researchers actively participate in the daily lives of the participants, observing and interacting with them in their natural settings.
- Thick Description: Ethnography aims to provide thick descriptions that capture the rich cultural context and the meanings attributed to behaviors and practices.
- Long-Term Engagement: Ethnographic research often involves extended periods of fieldwork to develop trust and rapport with the participants.

8.3.5 Content Analysis Research Method

Content analysis is a quantitative or qualitative research method that involves systematically analyzing the content of texts, documents, media, or other communication materials. It is commonly used to study media representations, messages, or patterns in communication.

Key Characteristics:

- Coding and Categorization: Researchers use coding schemes to categorize and analyze the content of the materials being studied.
- Objective and Replicable: Content analysis aims to be objective and replicable, ensuring that multiple researchers can analyze the same content and achieve similar results.

The selection of the appropriate research methods is crucial for answering the research questions and achieving the research objectives effectively. Researchers need to justify their chosen methods based on the research approach and the nature of the research problem. Combining multiple research methods in a mixed methods approach can provide a more comprehensive understanding of the research topic and enhance the validity and reliability of the findings. Careful consideration and thoughtful planning of the research methods ensure that the study is well-designed, rigorously executed, and capable of generating valuable insights for the intended audience.

8.4 Data Collection Tools

Discuss the data collection tools that will be used to gather information from participants or sources. This may include interviews, observations, focus group discussions, or document analysis. Explain why these tools are appropriate for the study and how they will help in answering the research questions.

The data collection tools and techniques used in a research proposal play a pivotal role in gathering relevant and reliable data to address the research questions effectively. These tools are essential for collecting information directly from participants or sources and help researchers gain insights into the research problem. Depending on the research approach (quantitative, qualitative, or mixed), researchers choose appropriate data collection methods to obtain the necessary data for analysis. Here's a more detailed explanation of the data collection tools commonly used in research proposals:

8.4.1 Interviews

Interviews are a widely used data collection tool in both qualitative and mixed-methods research. They involve direct communication between the researcher and the participants to gather detailed information about their experiences, opinions, attitudes, or behaviors. Interviews can be structured, semi-structured, or unstructured, depending on the level of flexibility in the questioning process.

- In-depth Interviews: Conducted one-on-one, in-depth interviews allow participants to share their experiences openly, providing rich and nuanced data.
- Focus Group Interviews: Focus groups involve multiple participants discussing a specific topic or issue, fostering group interactions and generating diverse perspectives.

8.4.2 Observations

Observations are valuable data collection tools in qualitative and mixed methods research. Researchers observe and record participants' behaviors, interactions, or activities in their natural settings. Observations can be participant observations, where the researcher actively participates, or non-participant observations, where the researcher remains an observer.

- Structured Observations: Researchers follow a predetermined observation protocol to focus on specific behaviors or events.
- Unstructured Observations: Observers take comprehensive notes on various aspects without a predefined checklist.

8.4.3 In-depth Case Interviews

In-depth case interviews are used primarily in qualitative research to study specific cases or individuals in detail. Researchers conduct extensive interviews to gain a comprehensive understanding of the case's complexities and context.

8.4.4 Focus Group Discussions (FGDs)

FGDs involve small groups of participants engaged in open discussions about specific topics or issues. They allow researchers to explore group dynamics, social norms, and shared perspectives.

8.4.5 Key Informant Interviews (KIIs)

Key informant interviews target individuals who possess specialized knowledge or expertise related to the research topic. These interviews provide valuable insights and expert opinions.

8.4.6 Participatory Rural Appraisal (PRA)

PRA is a qualitative research tool that involves the active involvement of community members in data collection and analysis. It aims to empower the community and gain their perspectives on development issues.

8.4.7 Discourse Analysis

Discourse analysis is a method used to analyze written or spoken language, texts, or conversations. It helps uncover underlying meanings, power relations, and ideological influences.

8.4.8 Selecting Data Collection Tools

- Choose data collection tools that align with the research approach and research questions. Ensure they capture the necessary data to address the research objectives effectively.
- Consider the context and cultural sensitivity when selecting data collection methods, especially in cross-cultural or ethnographic research.
- Conduct a pilot study or pretesting of data collection tools to identify any shortcomings and refine the instruments for better data collection.

The data collected through these tools form the foundation of the research findings and conclusions. Researchers must carefully design and implement the data collection process to ensure the accuracy, validity, and reliability of the data. Ethical considerations, such as informed consent and confidentiality, must be upheld throughout the data collection phase to protect the rights and privacy of the participants. Well-chosen and effectively implemented data collection tools contribute to the overall rigor and credibility of the research proposal.

8.5 Data Collection Instruments Development

Describe the development process of data collection instruments, such as interview guides, schedules, or checklists (Islam, 2022b). Explain how the instruments will be pre-tested and refined to ensure their validity and reliability. Emphasize the need for clarity and consistency in the instruments. The development of data collection instruments is a crucial step in the research process, as these instruments serve as the tools through which researchers gather data from participants or sources. Data collection instruments help ensure that the research objectives are met and that the data collected are relevant, reliable, and valid for addressing the research questions effectively. Depending on the nature of the study and the chosen data collection methods, researchers develop various types of instruments such as schedules, guidelines, or checklists. Here's a more detailed explanation of data collection instruments development:

8.5.1 Schedules

Schedules are structured data collection instruments used in quantitative research, especially in surveys and experimental studies. They consist of a series of predetermined questions with predefined response options. Schedules ensure standardization in data collection, enabling researchers to quantify responses for statistical analysis.

- Questionnaire Schedules: Used in surveys, these schedules consist of closed-ended questions with fixed response options, making data entry and analysis more manageable.
- Experimental Schedules: In experimental studies, schedules outline the step-by-step procedures to be followed by the researcher during data collection, treatment, and measurement.
- Semi-structured schedule to collect mixed data.

8.5.2 Guidelines

Guidelines are used in qualitative research methods, such as interviews and focus group discussions. They are flexible and provide general themes or topics to guide the data collection process. Unlike structured schedules, guidelines offer researchers the freedom to explore participants' responses in more depth.

> Interview Guidelines: Used in qualitative interviews, these provide open-ended questions or topics to be covered during the interview. The interviewer has the flexibility to probe and follow up on participants' responses.
>
> Focus Group Discussion Guidelines: Guidelines for focus group discussions outline the key themes or questions to be discussed during the session, facilitating group interactions and generating rich data.

8.5.3 Checklists

Checklists are systematic lists used to ensure that specific information or observations are recorded consistently. They are often used in observational research, where researchers systematically note the presence or absence of particular behaviors or events.

> Observation Checklists: Used in observational studies, these checklists guide researchers to record specific behaviors or events as they occur in the natural setting.
>
> Content Analysis Checklists: For content analysis, checklists help categorize and code specific features or elements in texts or media.

8.5.4 Development Process

> Clearly Define Objectives: Before designing data collection instruments, clearly define the research objectives and the specific information needed to address the research questions.

- Literature Review: Conduct a thorough literature review to identify relevant questions, themes, or factors that should be included in the data collection instruments.
- Pilot Testing: Before finalizing the instruments, conduct a pilot test with a small group of participants to identify any ambiguities or shortcomings in the questions or guidelines. This allows for refinements and improvements to ensure clarity and effectiveness.
- Ensure Clarity and Unbiased Language: Phrasing questions in clear, unambiguous language is essential to avoid confusion and ensure that participants interpret them as intended. Language should also be unbiased to avoid leading or influencing responses.
- Consider Cultural Sensitivity: In cross-cultural research, consider cultural norms and sensitivity when designing data collection instruments to ensure the questions are relevant and appropriate for the participants.
- Pre-Test for Reliability and Validity: Assess the reliability and validity of the instruments to ensure that they consistently measure what they are intended to measure.

Well-designed data collection instruments contribute to the success of the research project by providing accurate and relevant data to answer the research questions. A systematic and rigorous approach to instrument development enhances the credibility and trustworthiness of the research findings. By carefully tailoring the instruments to the research objectives and chosen data collection methods, researchers can ensure that the collected data are of high quality and contribute to a comprehensive and valuable research outcome.

8.6 Sampling and Respondents

8.6.1 Sampling

Outline the sampling strategy that will be used to select participants or sources for data collection. Explain the rationale behind the chosen sampling method and how it aligns with the research objectives. Clearly define the target population and describe how respondents will be recruited. Sampling and respondents are critical components of the research methodology in a research proposal. They involve the process of selecting a subset of individuals or units (the sample) from a larger population to represent and generalize findings to the entire population. The choice of the sampling method and the selection of appropriate respondents significantly impact the validity, generalizability, and reliability of the research findings. Sampling is the process of selecting a subset of individuals, units, or elements from a larger population for data collection. It is often not feasible or practical to study an entire population, so researchers use sampling techniques to conclude the entire group based on the characteristics of the selected sample.

Types of Sampling Techniques:

Probability Sampling: In probability sampling, each member of the population has a known and non-zero chance of being selected in the sample. It ensures that the sample is representative of the population and allows for statistical inference.

Simple Random Sampling: Every individual in the population has an equal chance of being selected.

Stratified Sampling: The population is divided into subgroups (strata), and random samples are taken from each stratum.

Systematic Sampling: Researchers select every nth member from a list of the population.

Cluster Sampling: The population is divided into clusters, and a random sample of clusters is selected.

Non-Probability Sampling: In non-probability sampling, the probability of each member being selected is unknown or unequal. Non-probability sampling is often used when probability sampling is not feasible or practical.

Convenience Sampling: Researchers select participants based on their accessibility and convenience.

Snowball Sampling: Participants refer other potential participants, creating a chain or "snowball" effect.

Purposive Sampling: Researchers deliberately select participants who meet specific criteria relevant to the research.

Quota Sampling: Researchers select participants to fulfill predetermined quotas based on specific characteristics.

8.6.2 Respondents

Respondents refer to the individuals, participants, or units from whom data is collected during the research process. They are the key sources of information, and their involvement is critical for obtaining insights into the research problem.

Considerations for Selecting Respondents:

- Relevance to Research Objectives: Ensure that the selected respondents are relevant to the research objectives and the research questions being addressed.
- Inclusion and Exclusion Criteria: Define clear criteria for inclusion and exclusion of respondents based on specific characteristics or attributes.
- Representation: Strive for diversity and representation in the sample to ensure that the findings can be generalized to the larger population.
- Accessibility and Feasibility: Consider the practicality and accessibility of the respondents, especially in terms of data collection logistics and resources.
- Informed Consent: Obtain informed consent from the respondents, ensuring that they are aware of the research purpose, their role, and any potential risks involved.
- Anonymity and Confidentiality: Assure respondents of the confidentiality and anonymity of their responses to encourage open and honest participation.

Selecting an appropriate sample and recruiting suitable respondents is crucial for the success of the research study. A well-chosen sample and respondents enhance the external validity of the findings, allowing researchers to make meaningful inferences about the larger population. Additionally, a transparent and justifiable sampling method helps strengthen the credibility of the research proposal and ensures the research is conducted in an ethical and scientifically rigorous manner.

8.7 Data Measurement

Detail the variables and measures that will be used to collect data. Describe the operationalization of variables and how they will be quantified or described. Discuss any instruments or scales used for data measurement and explain their validity and reliability. Data measurement and data analysis techniques are essential components of the research methodology in a research proposal. Data measurement involves defining and operationalizing variables to quantify or describe the data collected during the research process. Data analysis techniques, on the other hand, refer to the methods used to analyze the collected data and derive meaningful insights. These aspects are crucial in ensuring that the research findings are valid, reliable, and accurately address the research questions. Here's a more detailed explanation of data measurement and data analysis techniques:

Defining Variables: Defining variables involves specifying the concepts or characteristics that will be measured during the research. Variables can be quantitative (numeric) or qualitative (categorical). Researchers must clearly articulate the variables and their operational definitions to ensure consistency and accuracy in data collection.

Operationalization: Operationalization refers to the process of translating abstract concepts into measurable indicators or variables. It involves defining how variables will be measured or observed to obtain numerical or categorical data. Researchers need to select appropriate indicators or measurement scales that align with the research objectives and the type of data needed.

8.7.1 Types of Data Measurement

Quantitative Data Measurement: In quantitative research, variables are typically measured using standardized instruments, surveys, or scales that yield numerical data. Common measurement scales include Likert scales, rating scales, and interval or ratio scales.

Qualitative Data Measurement: Qualitative research measures variables using descriptive and non-numeric data, such as narratives, textual responses, or observations. Researchers categorize and interpret qualitative data to derive meaningful themes or patterns.

8.8 Data Analysis Techniques

8.8.1 Quantitative Data Analysis

Quantitative data analysis involves the use of statistical techniques to analyze numerical data collected during the research. The choice of analysis method depends on the research questions and the level of measurement of the variables.

Common Quantitative Data Analysis Techniques:

Descriptive Statistics: Summarizing and describing data using measures such as mean, median, mode, and standard deviation.

Inferential Statistics: Making inferences or generalizations about a population based on the sample data, using techniques like t-tests, ANOVA, and regression analysis.

Correlation Analysis: Assessing the relationship between variables using correlation coefficients.

Factor Analysis: Identifying underlying factors that explain patterns of observed variables.

Chi-Square Test: Analyzing the association between categorical variables.

8.8.2 Qualitative Data Analysis

Qualitative data analysis involves systematically interpreting non-numerical data to identify themes, patterns, and meanings. Researchers employ various methods to analyze qualitative data, ensuring rigor and credibility in their findings.

8.8.3 Common Qualitative Data Analysis Techniques

> Thematic Analysis: Identifying and analyzing recurring themes or patterns in the data.
> Content Analysis: Analyzing the content of textual or visual data to identify key concepts or categories.
> Narrative Analysis: Analyzing participants' narratives or stories to gain insights into their experiences.
> Grounded Theory: Developing theoretical frameworks from the data itself, allowing new concepts and theories to emerge.

8.9 Data Validity and Reliability

Data validity refers to the accuracy and appropriateness of the data collected, while data reliability refers to the consistency and stability of the measurements. Researchers must consider both validity and reliability to ensure that the data accurately represent the research constructs and are suitable for drawing meaningful conclusions.

By carefully selecting appropriate data measurement techniques and data analysis methods, researchers can draw robust conclusions and make meaningful interpretations from the collected data. Transparently explaining these aspects in the research proposal enhances its methodological rigor and reinforces the credibility of the study's findings. Proper data measurement and analysis contribute significantly to the overall success and impact of the research project. Explain the data analysis techniques that will be applied to analyze the collected data. For quantitative research, specify the statistical methods that will be used. For qualitative research, describe how thematic analysis or content analysis will be conducted. Justify the selected analysis techniques based on the research questions.

8.10 Limitations of the Study

Acknowledge the limitations of the research design and methodology. Discuss potential challenges and constraints that may impact the validity or generalizability of the findings. Being transparent about the limitations demonstrates the researcher's awareness of the study's boundaries. Several difficulties can arise during the data collection process in the field, but this edited compilation strongly asserts that the primary and foremost factor determining these challenges is the local situation. Researchers should accord significant attention to this aspect (Islam et al., 2021).

Addressing the limitations of a study is a crucial aspect of a research proposal. Limitations refer to the potential weaknesses, shortcomings, or constraints that may affect the research design, data collection, analysis, and interpretation of results. Every research study, regardless of its rigor, is bound to have limitations, and researchers need to be transparent about these limitations in their proposal. Acknowledging and discussing limitations demonstrates the researcher's awareness of potential challenges and strengthens the credibility and validity of the research findings. Here's a more detailed explanation of the limitations of the study in a research proposal:

Scope and Generalizability: One of the common limitations of a research study is the scope of the research. Studies are often conducted with a specific focus, sample size, or time frame, which may limit the generalizability of the findings to a broader population or different contexts. Researchers should explicitly state the scope of their study and acknowledge that the results may not be applicable beyond the selected sample or settings.

Sample Size and Representativeness: The sample size chosen for the study may be limited by practical considerations, budget constraints, or time limitations. Smaller sample size can affect the statistical power of the study and may limit the ability to detect significant relationships or differences. Additionally, researchers may face challenges in obtaining a truly representative sample, especially when studying hard-to-reach populations or vulnerable groups.

Data Collection and Measurement: Limitations can arise from the data collection methods used in the study. For instance, self-reported data may be subject to recall bias or social desirability bias, affecting the accuracy of responses. Researchers should acknowledge any potential limitations in data collection techniques and discuss how they mitigated these issues.

Research Design and Methodology: The research design and methodology selected for the study may have inherent limitations. For example, cross-sectional studies can only provide a snapshot of data at a particular point in time, limiting the ability to establish causal relationships. Longitudinal studies may face attrition and participant dropouts over time, which can affect the validity of the findings.

External Factors and Confounding Variables: External factors and confounding variables are variables that may influence the research outcomes but are not directly measured or controlled by the researcher. Researchers should acknowledge the potential influence of these variables and discuss how they attempted to control for them or their impact on the results.

Data Analysis and Interpretation: Data analysis techniques may also have limitations. The choice of statistical tests, assumptions made, or the exclusion of certain data points can impact the results. Researchers should be transparent about the analytical methods used and the potential impact on the findings.

Time and Resource Constraints: Time and resource constraints are common limitations in research. Researchers may face limitations in terms of funding, manpower, or access to data. These constraints can affect the depth and scope of the study.

Unforeseen Circumstances: Despite careful planning, researchers may encounter unforeseen circumstances during the research process that can affect the study's outcomes. These could include changes in the external environment, disruptions during data collection, or other unforeseen events.

Bias and Subjectivity: Researchers must be aware of their own biases and subjectivity that may influence the study design, data collection, and analysis. Transparently addressing potential biases and strategies to minimize them enhances the research's credibility.

Recommendations for Future Research: In the limitations section, researchers can offer insights and recommendations for future research to address the identified limitations. These recommendations can guide other researchers in building on the current study and expanding the knowledge base in the field. Recognizing and discussing the limitations of the study in the research proposal is not a sign of weakness but rather an indication of the researcher's intellectual honesty and dedication to rigorous scientific inquiry. By being transparent about potential challenges and limitations, researchers demonstrate a commitment to improving the quality of research and contribute to a deeper understanding of the study's results. Moreover, identifying limitations can open avenues for future research that can address the gaps and strengthen the overall body of knowledge in the field.

8.11 An Example of the Step-by-Step Methodology of a Research Proposal

Research Proposal Title: "Exploring Social Media's Impact on Mental Health: A Qualitative Study in Dhaka City, Bangladesh"

Location of the Study: This research will be conducted in urban and suburban areas of Dhaka City, where high social media usage is prevalent. The diverse settings will provide a rich context to understand the varied experiences of individuals.

Research Approach: A qualitative research approach will be employed to delve into the subjective experiences and perceptions of individuals regarding the impact of social media on mental health.

Research Methods: In-depth interviews and focus group discussions will be the primary methods, allowing for a nuanced exploration of participants' attitudes, emotions, and behaviors related to social media use and its effects on mental well-being.

Data Collection Tools: The research will utilize semi-structured interview guides and focus group discussion protocols to ensure consistency in the data collection process while allowing flexibility for participants to share their unique perspectives.

Data Collection Instruments Development: The interview and discussion instruments will be developed through an iterative process, incorporating feedback from experts in psychology, social sciences, and individuals with experience in studying social media impacts.

Sampling and Respondents: A purposive sampling strategy will be employed to select participants with diverse social media usage patterns, ages, and socio-economic backgrounds. This approach aims to capture a broad range of experiences.

Data Measurement and Data Analysis Techniques: Qualitative data will be measured through thematic analysis, identifying recurring themes and patterns in participants' narratives to draw meaningful insights. Thematic analysis will be complemented by constant comparative analysis to systematically compare and contrast emerging themes, ensuring rigor and reliability in the analysis process.

Data Validity and Reliability: To enhance validity, member-checking sessions will be conducted, allowing participants to verify the accuracy of their contributions. Reliability will be ensured through inter-coder reliability checks in the analysis phase.

Limitations of the Study: Limitations include the potential for social desirability bias, as participants may provide responses deemed socially acceptable. Additionally, the qualitative nature of the study limits generalizability, emphasizing the need for cautious interpretation of findings.

CHAPTER 9

Research Ethics of a Research Proposal

ABSTRACT
This chapter thoroughly examines the paramount importance of research ethics in the development of a research proposal. Ethics in research serves as a cornerstone, ensuring the responsible and respectful conduct of studies involving human subjects, data, and diverse methodologies. Moreover, the chapter comprehensively explores the ethical considerations researchers must navigate, encompassing informed consent, confidentiality, data integrity, and participant welfare. A nuanced discussion on the ethical challenges posed by emerging technologies, cross-cultural studies, and sensitive topics adds depth to the exploration. Furthermore, practical guidance is offered on navigating ethical review processes and aligning proposals with established ethical guidelines. As research increasingly intersects with societal values, this chapter provides an essential guide for researchers to uphold the highest ethical standards in their pursuit of knowledge.

KEYWORDS: Research ethics, informed consent. confidentiality, participant eelfare, ethical review process

9.1 Meaning of Research Ethics and Its Importance

9.1.1 Research Ethics

Research ethics refers to the principles, guidelines, and standards that govern the conduct of research involving human participants, animals, or the collection and analysis of data. It is a set of moral principles and rules of conduct that guide researchers in ensuring the integrity, honesty, and responsible conduct of their studies. Research ethics encompasses a range of considerations, including the rights and well-being of participants, the veracity of research findings, and the societal implications of research.

9.1.2 Importance of Research Ethics

Research ethics is one of the most important aspects of conducting any research at the community level that investigates human behavior. When seeking legitimate and informed consent, especially within local contexts, it is crucial to consider local ethical norms and linguistic nuances. In certain political and social environments, participants need to be informed about potential risks associated with their involvement. In these local perspectives, informed consent holds significant value. Participants should receive comprehensive information about the research, including its purpose, funding sources, the nature of their involvement, how findings will be utilized, potential benefits, and the future use of collected data. It is imperative to communicate any potential risks that participants might face due to their involvement, especially when working with minority or marginalized groups in politically hostile environments. Clear disclosure should be made regarding any foreseeable limitations on anonymity and confidentiality. Researchers bear the responsibility of ensuring that participants understand the legal status of anonymity and confidentiality in their country and the circumstances under which the researcher might be compelled to disclose information (Islam et al., 2014). Without adhering to proper ethical principles, challenges in research ethics arise, contributing to increased complexity in the research process. Islam and Hajar (2013) have categorized the challenges into two main groups: commune-level challenges and procedural-level challenges. Commune-level challenges are closely linked to the existing conditions of the community and encompass factors such as low socio-economic conditions, cultural barriers, limited research knowledge, and a lack of cooperation from both funding and operating organizations. On the other hand, procedural level challenges pertain to the limitations inherent in the qualitative research approach during the data collection process. Noting that cultural diversity is a reality, we would assert that this occurrence poses a challenge for cross-cultural or intercultural research (Islam & Banda, 2005). These challenges involve concerns about the reliability and validity of research findings, the complexity and diversification of human behavior, adherence to research ethics, the unavailability and inaccessibility of data, and power relations. It's important to note that these challenges are interconnected and mutually influence each other, contributing to the overall complexity of the research process.

Protection of Participants:

- Human Dignity and Rights: Research ethics ensures that individuals participating in studies are treated with dignity and that their rights are respected. This includes the right to informed consent, privacy, and protection from harm.

Credibility of Research:

- Research Integrity: Ethical conduct is integral to the integrity of research. Adhering to ethical standards enhances the credibility and reliability of research findings, fostering trust among the research community and the public.

Avoidance of Harm:

- Non-Maleficence: Ethical guidelines emphasize the principle of non-maleficence, requiring researchers to minimize the potential for harm to participants. This involves careful consideration of the risks and benefits of the research.

Informed Decision-Making:

- Informed Consent: Research ethics promotes the concept of informed consent, ensuring that participants have a clear understanding of the research purpose, procedures, and potential risks. Informed participants can make autonomous decisions about their participation.

Confidentiality and Privacy:

- Data Protection: Ethical standards mandate the protection of participants' confidentiality and privacy. Researchers must implement measures to safeguard sensitive information and ensure that participants' identities remain confidential.

Transparency and Reproducibility:

- Open Communication: Ethical research involves transparent communication about the research process. This includes providing detailed information about methodologies, data collection, and analysis methods, contributing to the reproducibility of research.

Social Responsibility:

- Societal Impact: Ethical considerations extend beyond individual studies to encompass the broader societal impact of research. Researchers are encouraged to reflect on the potential implications of their work on communities, cultures, and societal values.

Compliance with Regulations:

- Legal and Institutional Compliance: Adhering to research ethics ensures compliance with legal and institutional regulations. Many research institutions and funding agencies require researchers to undergo ethical review and approval before conducting studies.

Cross-Cultural Sensitivity:

- Respect for Diversity: In a globalized research landscape, ethical considerations include respect for cultural, social, and individual differences. Researchers must navigate diverse perspectives and ensure that their studies are culturally sensitive.

Emerging Technologies and Issues:

- Technological and Contemporary Challenges: Research ethics evolves to address ethical challenges posed by emerging technologies, such as artificial intelligence and genetic research. It also engages with contemporary issues, such as social justice and environmental sustainability.

9.2 Ethical Principles and Guidelines

Ethical issues in a research proposal are of paramount importance as they pertain to the ethical principles and guidelines that must be followed when conducting research involving human participants or sensitive data. Several considerations must be taken into account by a researcher before commencing data collection. These encompass obtaining ethical clearance from the relevant ethical body or committee, securing informed consent, establishing contracts with various stakeholders, configuring field settings, allocating and managing time effectively, leading the field team, addressing contextual and cultural diversities, navigating community settings, understanding the socio-economic and psychological patterns of the community, analyzing political dynamics, fostering rapport between data collectors and respondents, obtaining permission to access the community, determining the language and mode of data collection, acknowledging power relations, recognizing the role of gatekeepers, managing privacy and confidentiality issues, understanding layers of expectations among researchers, respondents, and funding organizations, and deciding on the method of data recording, whether written, through memorization, voice recording, or video recording (Islam et al., 2021). Addressing ethical considerations ensures the protection of participants' rights, welfare, and confidentiality, as well as the integrity of the research process. Ethical considerations are critical in building trust with participants, the research community, and society at large. Here's a more detailed explanation of ethical principles and guidelines in a research proposal:

9.2.1 Informed Consent

Informed consent is a cornerstone of ethical research involving human participants. Researchers must obtain voluntary, informed, and written consent from all participants before they are included in the study. Informed consent ensures that participants are fully aware of the research purpose, procedures, potential risks, benefits, and their right to withdraw from the study at any time without repercussions.

9.2.2 Confidentiality and Anonymity

Researchers must guarantee confidentiality by safeguarding participants' personal information and ensuring that their identities remain anonymous in research reports. Participants' names and identifying information should not be disclosed without their explicit consent. Anonymity is particularly critical in sensitive research areas to protect participants from potential social, legal, or professional consequences.

9.2.3 Privacy and Data Protection

Researchers should prioritize participants' privacy and protect their data. All data collected should be securely stored and accessed only by authorized personnel. Compliance with relevant data protection regulations, such as GDPR (General Data Protection Regulation), is essential when handling personal data.

9.2.4 Minimizing Harm and Risk

Researchers must strive to minimize any physical, psychological, social, or economic harm that participants may encounter during the research process. A comprehensive risk assessment should be conducted to identify potential risks, and appropriate measures should be put in place to mitigate them.

9.2.5 Beneficence and Non-maleficence

Researchers should act in the best interest of participants (beneficence) and avoid causing harm (non-maleficence). The potential benefits of the research should outweigh any potential risks to participants.

9.2.6 Fair Selection of Participants

The selection of research participants should be fair and just, avoiding any form of discrimination or bias. Researchers should avoid targeting vulnerable populations solely because of their perceived accessibility or susceptibility.

9.2.7 Collaboration and Community Involvement

In research involving specific communities, it is essential to collaborate with community representatives and involve them in the research process. Community engagement helps ensure that research aligns with the community's needs, values, and interests.

9.2.8 Ethical Review and Approval

Researchers must seek ethical review and approval from an institutional review board (IRB) or ethics committee before commencing the study. The review process ensures that the research proposal adheres to ethical standards and guidelines.

9.2.9 Transparency and Full Disclosure

Researchers should provide clear and transparent information about the research aims, procedures, funding sources, and potential conflicts of interest. Full disclosure helps build trust with participants and stakeholders.

9.2.10 Reporting and Dissemination

Researchers should report the research findings accurately and responsibly, ensuring that the results are not misrepresented or exaggerated.

Negative or inconclusive results should also be reported to avoid publication bias.

9.3 Informed Consent Procedures

Information Disclosure: Researchers are responsible for providing participants with clear and understandable information about the research. This includes the purpose of the study, procedures involved, potential risks and benefits, and the voluntary nature of participation.

Comprehension and Voluntariness: Participants must comprehend the information provided before giving consent. Researchers should use layman's terms and encourage questions to ensure understanding. Moreover, participants should feel free to decline or withdraw from the study at any time without repercussions.

Documentation: Informed consent is typically documented through a consent form. The form should include all relevant details about the study and be written in a language accessible to participants. Participants are often given a copy of the signed consent form for their records.

Special Considerations: For vulnerable populations, such as minors, individuals with cognitive impairments, or those with limited decision-making capacity, additional precautions are taken. In such cases, researchers may seek assent from the participant and consent from a legal guardian.

Ongoing Consent: Consent is not a one-time event; it is an ongoing process. Researchers should maintain open communication with participants throughout the study, updating them on any changes to the research or newly identified risks.

Withdrawal and Debriefing: Participants have the right to withdraw from the study at any stage without facing negative consequences. Researchers should provide a clear process for withdrawal and offer debriefing sessions to address any lingering concerns.

9.4 Confidentiality and Privacy

Data Protection: Researchers must implement measures to protect the confidentiality of participants' data. This includes secure storage, restricted access, and encryption where applicable.

> Anonymity vs. Confidentiality: While anonymity implies that the participant's identity is unknown even to the researcher, confidentiality involves protecting the disclosed information. Researchers must communicate whether data will be anonymous or confidential.
>
> Informed Consent: The consent process should explicitly address how participants' confidentiality will be maintained. Researchers should inform participants about any potential exceptions, such as legal requirements to report certain information.
>
> Limits of Confidentiality: Researchers should be transparent about the limits of confidentiality. If there are circumstances under which confidentiality may be breached (e.g. risk of harm to self or others), these should be communicated during the consent process

9.5 Risk Assessment and Mitigation

> Types of Risks: Researchers must identify and categorize potential risks associated with the research. These can include physical, psychological, social, legal, and informational risks to participants.
>
> Risk Assessment Criteria: Develop criteria for assessing the severity, likelihood, and significance of identified risks. This allows for a systematic evaluation of the potential impact of each risk on participants and the research process.

- Informed Consent and Risk Disclosure: Communicate identified risks to participants during the informed consent process. Provide detailed information about the nature and potential consequences of risks to enable participants to make informed decisions about their participation.
- Continuous Monitoring: Establish mechanisms for continuous monitoring of risks throughout the research lifecycle. This includes regular check-ins with participants, ongoing assessment of potential emerging risks, and adjustments to research procedures as needed.
- Participant Welfare: Prioritize participant welfare in the research design. Mitigation strategies should focus on minimizing potential harm and ensuring the well-being of participants throughout the study.

9.6 Research with Vulnerable Populations

- Definition of Vulnerable Populations: Identify and define the specific vulnerable populations involved in the research. This may include minors, individuals with cognitive impairments, economically disadvantaged groups, refugees, and others who may face heightened risks.
- Informed Assent and Consent: In addition to obtaining informed consent from adult participants, researchers should seek assent from minors and, when applicable, consent from legal guardians. The consent process should be adapted to the cognitive abilities and comprehension levels of the participants.
- Ethical Review Board Scrutiny: Studies involving vulnerable populations undergo heightened scrutiny by ethical review boards. Researchers must provide robust justifications for the inclusion of vulnerable participants and demonstrate comprehensive ethical safeguards.

Risk-Benefit Ratio: Conduct a thorough assessment of the risk-benefit ratio, considering the potential benefits of the research against the potential harms to vulnerable participants. The welfare and best interests of the participants should be paramount.

9.7 Data Management and Integrity

Data Collection Protocols: Establishing clear and standardized protocols for data collection is a fundamental step in ensuring the reliability and validity of research findings. These protocols serve as a blueprint, delineating the specific procedures, tools, and methodologies employed during the data collection phase. A well-defined protocol promotes consistency across data collection efforts, reducing variability and enhancing the overall reliability of the research. Researchers should meticulously outline the step-by-step processes involved, from participant recruitment and data acquisition to storage and analysis. Additionally, the protocol should address potential challenges and provide guidelines for maintaining data quality throughout the research endeavor.

Data Security: The implementation of robust data security measures is imperative to safeguard research data from unauthorized access, loss, or corruption. Researchers must establish secure storage systems that meet industry standards for data protection. This involves employing encryption techniques to secure data during transmission and storage. Access to research data should be restricted to authorized personnel only, with well-defined access controls and user authentication mechanisms in place. Regular monitoring and audits of data security protocols are essential to identify and address potential vulnerabilities, ensuring the confidentiality and integrity of the data throughout its lifecycle.

Researcher Training: Ensuring that researchers and research staff are well-versed in data management best practices is crucial for maintaining the integrity of the research process. Adequate training programs should cover a spectrum of topics, including the intricacies of various data collection methods, protocols for secure data storage, and ethical considerations related to data handling. Researchers should be equipped with the skills to implement data collection protocols effectively, use data management tools proficiently, and navigate ethical dilemmas that may arise during the research journey. Ongoing training initiatives contribute to a research team's collective competence in data management and ethical research practices.

Data Transparency: Emphasizing transparency in data management is essential for enhancing the reproducibility and verifiability of research findings. Transparent practices involve clear and comprehensive documentation of the entire data lifecycle, from the conceptualization of data collection procedures to the dissemination of results. Researchers should provide detailed descriptions of data collection instruments, sampling methodologies, and any modifications made to the original protocols. Transparent reporting of data analysis methods and statistical procedures further enhances the credibility of the research. By openly sharing the processes involved in data collection, storage, and analysis, researchers contribute to the advancement of scientific knowledge and foster a culture of accountability in research practices.

9.8 Publication Ethics

Authorship and Contributorship: Authorship and contributorship guidelines are paramount in ensuring transparency and fairness in acknowledging individuals' contributions to a research project. It is essential to establish clear criteria for authorship within the research team, outlining the specific contributions that qualify an individual as an author. These contributions may encompass the conception and design of the study, data collection, analysis, and interpretation. Institutions and research teams should develop policies to address authorship disputes and provide a framework for acknowledging the diverse roles of team members.

- Corresponding Author Responsibility: Clearly define the responsibilities of the corresponding author, who often serves as the primary point of contact with the journal or publisher. This includes overseeing the submission process, responding to queries, and managing post-publication communication.
- Acknowledgment of Contributions: Acknowledge non-author contributors appropriately. Individuals who made valuable contributions but did not meet the criteria for authorship should be recognized in the acknowledgment section of the publication.
- Authorship Statement: Require authors to provide a detailed authorship statement that specifies the contributions of each author. This statement can be included in the manuscript or as a supplementary document.
- Collaborative Research Agreements: Establish collaborative research agreements at the beginning of a project. Clearly outline expectations for authorship, contributorship, and the process for resolving disputes. This proactive approach can prevent conflicts during the later stages of the research.

Plagiarism Prevention: Plagiarism is a serious ethical breach that undermines the integrity of research. Researchers must adopt measures to prevent plagiarism and uphold the originality of their work. This involves more than just avoiding verbatim copying; it extends to proper citation practices, paraphrasing, and the use of plagiarism detection tools.

- Self-Plagiarism Awareness: Researchers should be vigilant about self-plagiarism, which involves reusing one's own previously published work without proper citation. Journals and publishers often have policies regarding self-plagiarism that authors must adhere to.

- Proactive Citation Practices: Encourage proactive citation of relevant sources throughout the manuscript. This not only avoids plagiarism but also enriches the context of the research by acknowledging the existing body of knowledge.
- Citation Management Tools: Familiarize researchers with citation management tools that facilitate proper referencing. These tools can help organize references, ensure consistency, and generate citation lists in various formats.
- Educational Initiatives: Implement educational initiatives on plagiarism prevention. Workshops, training sessions, or online resources can raise awareness among researchers about the nuances of plagiarism and how to avoid unintentional breaches.

Peer Review Integrity: Maintaining the integrity of the peer review process is essential for the quality and credibility of published research. Peer reviewers play a crucial role in evaluating the rigor, validity, and significance of research submissions.

- Blind Peer Review: Consider adopting a blind peer review process where the identities of authors and reviewers are concealed. This helps mitigate biases and ensures that the evaluation is based solely on the merit of the research.
- Conflict of Interest Disclosure: Require peer reviewers to disclose any potential conflicts of interest that could compromise their objectivity. This transparency enhances the credibility of the peer review process.
- Timely and Constructive Feedback: Emphasize the importance of providing timely and constructive feedback. Peer reviewers should focus on the strengths and weaknesses of the research, offering suggestions for improvement.
- Editorial Oversight: Editors should exercise editorial oversight to ensure the fairness and quality of the peer review process. This includes selecting competent reviewers and monitoring the feedback provided.

Data Integrity and Reproducibility: Ensuring the integrity and reproducibility of research data is fundamental to the scientific process.

Researchers should adopt practices that promote transparency and facilitate the validation of results through access to raw data and methodologies.

- Data Sharing Platforms: Encourage researchers to deposit raw data in reputable data-sharing platforms or repositories. This not only promotes transparency but also allows other researchers to verify and build upon the findings.
- Comprehensive Methodology Section: Include a detailed methodology section in the manuscript that outlines the procedures followed during data collection and analysis. Clarity in this section aids in understanding the research process and enhances reproducibility.
- Data Availability Statements: Require authors to provide data availability statements indicating where the data can be accessed. This statement communicates the researcher's commitment to transparency and data sharing.
- Validation and Verification Protocols: Describe validation and verification protocols for data accuracy. This may involve detailing quality control measures, calibration procedures, and steps taken to address potential biases in the data.
- Documentation of Analytical Methods: Document the analytical methods used in data analysis. This includes software used, parameter settings, and any modifications made during the analysis process.

9.9 Conflict of Interest

Disclosure of Financial Interests: Researchers must transparently disclose any financial interests or affiliations that could be perceived as a conflict of interest. This includes financial relationships with industry, funding sources, or organizations that may stand to gain from the research outcomes.

Impartial Decision-Making: Uphold impartial decision-making by identifying and managing conflicts of interest throughout the research process. This is crucial in situations such as peer review, data analysis, and the interpretation of results.

Institutional Oversight: Institutions and research organizations should establish mechanisms for identifying, reviewing, and managing conflicts of interest. This may involve the creation of COI committees or designated individuals responsible for overseeing and addressing potential conflicts.

9.10 Ethical Oversight and Institutional Review Boards (IRBs)

Ethical oversight, facilitated by Institutional Review Boards (IRBs) or Ethics Review Committees, is a crucial component of research governance. This section explores the fundamental principles, functions, and best practices associated with ethical oversight in research, emphasizing the role of IRBs in safeguarding the rights and well-being of research participants.

9.10.1 Key Principles

Protection of Participants: The primary role of ethical oversight is to ensure the protection of human participants involved in research. This includes safeguarding their rights, welfare, and confidentiality throughout the research process.

Informed Consent: Ethical oversight emphasizes the importance of obtaining informed consent from research participants. This process involves providing clear, comprehensible information about the research, its objectives, potential risks, and benefits, allowing participants to make voluntary and informed decisions about their involvement.

Risk-Benefit Assessment: IRBs conduct a thorough risk-benefit assessment, weighing potential harms against anticipated benefits. Researchers must justify the necessity of the study, and IRBs play a critical role in evaluating whether the potential benefits outweigh the risks.

9.10.2 Functions of Institutional Review Boards

Protocol Review: IRBs review research protocols to ensure compliance with ethical standards. This includes assessing the study design, methodology, participant recruitment, and the measures in place to protect participants.

Informed Consent Review: IRBs scrutinize informed consent documents to verify clarity, completeness, and appropriateness. They ensure that participants are adequately informed about the research and their rights and that the consent process is ethically sound.

Risk Assessment: IRBs assess the potential risks associated with the research, considering physical, psychological, social, legal, and economic factors. They work to minimize these risks and ensure that researchers have plans in place to address adverse events.

Participant Recruitment and Selection: IRBs evaluate participant recruitment strategies to ensure they are fair, and transparent, and do not unduly influence individuals to participate. Additionally, they review criteria for participant selection to prevent unfair exclusions or biases.

Privacy and Confidentiality: Ensuring the privacy and confidentiality of research participants is a key focus. IRBs assess the measures in place to protect participant data and ensure that researchers adhere to strict confidentiality standards.

9.11 An Example of the Research Ethics of a Research Proposal

Research Proposal Title: "Ethical Considerations in Online Survey Research on Sensitive Topics: A Study in London City"

Researching sensitive topics requires a stringent ethical framework to safeguard the well-being and privacy of participants. This study, aiming to explore individuals' experiences with stigmatized health conditions through an online survey in London City, adheres to established ethical principles. Informed consent will be obtained from participants, clearly outlining the study's purpose, potential risks, and the voluntary nature of their involvement. Confidentiality will be maintained by using anonymized data, and participants will have the option to skip any questions they find uncomfortable. The research design aligns with ethical guidelines, ensuring that the knowledge gained contributes to the broader understanding of these sensitive issues while minimizing potential harm to participants. This study has received approval from the Institutional Review Board (please mention the name of the Ethical Board with the permission of reference number) to ensure adherence to ethical standards throughout the research process.

CHAPTER 10

Scope of the Research of a Research Proposal

ABSTRACT
This chapter provides an insightful exploration of the critical component known as the "scope of the research" within a research proposal. It begins by unraveling the essence of this section, which delineates the boundaries and parameters of the study. Moreover, the chapter navigates the nuances of determining the study's scope, encompassing the facets that will be included, as well as those that will be excluded. Practical guidance is offered on articulating a well-constructed "Scope of Research" section that encapsulates the research's focal points and limits. Furthermore, the chapter underscores the importance of maintaining a balanced scope that is neither overly ambitious nor excessively restrictive. Through a systematic approach, researchers are guided in crafting a "Scope of Research" that aligns with the research's objectives while being transparent about its constraints.

KEYWORDS: Scope of the research, boundaries, parameters, objectives, transparency

10.1 What Is the Scope of the Research in a Research Proposal?

The scope of the research in a research proposal refers to the extent and boundaries of the study. It defines the parameters within which the research will be conducted and outlines what the study will cover and what it will not cover. The scope sets the limits for the research, ensuring that the study remains focused, feasible, and manageable. It is an essential element of the research proposal as it helps both the researchers and the readers understand the scale and depth of the study.

10.2 What Will We Include in the Scope of the Study?

In the scope of study section of a research proposal, you should include specific details that define the boundaries and parameters of your research. This section helps readers understand what your study will cover and what it will not cover. Here are the key components to include in the scope of the study:

- Research Objectives and Questions: Clearly state the main research objectives and research questions that your study aims to address. This sets the foundation for the scope of the research.
- Geographic Scope: Define the geographical area or locations that will be included in the study. Specify the regions, countries, or communities that your research will focus on.
- Time Frame: Indicate the time that your study will cover. This includes the start and end dates of data collection or the timeframe during which historical data will be analyzed.
- Variables and Concepts: Identify the main variables and concepts that you will study. Explain which factors or elements you will measure and analyze in your research.
- Research Design and Methodology: Describe the type of research design you will use (e.g. qualitative, quantitative, and mixed methods) and the specific research methodologies you plan to employ (e.g. surveys, experiments, and case studies).
- Sample Size and Selection: Specify the approximate size of your sample and the criteria you will use to select participants or data points. Justify why this sample size and selection method are appropriate for your research.
- Data Collection Methods: Explain how you will collect data, such as through surveys, interviews, observations, or secondary sources. Detail the data collection tools or instruments you will use.
- Data Sources: Indicate the sources from which you will gather data. These could include primary sources (e.g. direct data

collection) or secondary sources (e.g. existing datasets and published literature).

Limitations and Delimitations: Acknowledge any limitations or constraints that may impact the research's scope, such as time constraints, budget limitations, or access to data. Also, mention any specific aspects intentionally excluded from the study (delimitations).

Practical and Ethical Considerations: Discuss practical considerations related to the feasibility of data collection, access to resources, or logistical challenges. Address ethical considerations, such as obtaining informed consent and ensuring participant confidentiality.

Significance of the Scope: Explain why the defined scope is appropriate and relevant for addressing your research objectives and questions. Emphasize the value and uniqueness of your study within this scope.

By including these components in the scope of the study section, you provide a clear understanding of the research's focus, limitations, and potential contributions. This helps reviewers and stakeholders evaluate the feasibility, significance, and relevance of your research proposal.

10.3 How to Write a Good "Scope of Research" Section?

Writing a good "Scope of Research" section in a research proposal involves clearly defining the boundaries and limitations of your study while also emphasizing its significance and relevance. Here are some guidelines to help you craft an effective 'Scope of Research' section:

Start with a Clear Statement: Begin the section with a concise and clear statement that outlines the overall scope of your research. Mention the main research objectives and the central research question(s) that your study aims to address.

- Define Geographic and Temporal Boundaries: Specify the geographical area or locations that your study will cover. Also, indicate the time frame during which data will be collected or analyzed. This provides context and context for the research.
- Identify the Study Variables: Clearly state the main variables and concepts that you will study. This helps readers understand the central focus of your research.
- Explain the Research Design and Methodology: Describe the type of research design you will use (e.g. quantitative, qualitative, and mixed methods) and the specific research methodologies you plan to employ (e.g. surveys, experiments, and case studies). Justify your choices based on the research objectives.
- Specify the Sample Size and Selection: Detail the approximate size of your sample and the criteria for selecting participants or data points. Justify why this sample size and selection method are appropriate for your research.
- Clarify Data Collection Methods: Explain how you will collect data, such as through surveys, interviews, observations, or secondary sources. Provide an overview of the data collection tools or instruments you will use.
- Identify Data Sources: Indicate the sources from which you will gather data. Mention whether you will use primary sources (e.g. direct data collection) or secondary sources (e.g. existing datasets and published literature).
- Address Limitations and Delimitations: Acknowledge any limitations or constraints that may impact the research's scope, such as time constraints, budget limitations, or access to data. Also, mention any specific aspects intentionally excluded from the study (delimitations).
- Discuss Practical and Ethical Considerations: Address practical considerations related to the feasibility of data collection, access to resources, or logistical challenges. Also, discuss ethical considerations, such as obtaining informed consent and ensuring participant confidentiality.

- Emphasize Significance and Relevance: Explain why the defined scope is appropriate and relevant for addressing your research objectives and questions. Highlight the value and uniqueness of your study within this scope.
- Link to Research Objectives: Ensure that the scope aligns closely with the research objectives and is in line with the central research question(s) you aim to answer.
- Be Concise and Cohesive: Keep the section concise and focused. Avoid unnecessary details and be cohesive in presenting the different components of the scope.

By following these guidelines, you can write a well-structured and persuasive "Scope of Research" section that effectively communicates the boundaries, relevance, and potential contributions of your research proposal. This section helps reviewers and stakeholders understand the scope and significance of your study and its alignment with the research objectives.

10.4 An Example of the Scope of the Research of a Research Proposal

Scope of the Study: "Digital Literacy and Social Inclusion in Canadian Rural Communities"

In the evolving landscape of technology, this study aims to explore the intersection of digital literacy and social inclusion within the unique context of rural communities in Canada. The scope of the research extends to understanding how access to and proficiency in digital technologies influence social interactions, community engagement, and overall well-being in rural settings across the country.

Geographically, the study encompasses diverse rural regions in Canada, acknowledging the varying degrees of digital infrastructure and connectivity. By including participants from different provinces, the research seeks to capture the distinct challenges and opportunities faced by rural communities in their quest for digital inclusion.

The temporal scope spans a defined period to observe changes and developments in digital literacy initiatives and their impact on social dynamics. Methodologically, a mixed-methods approach is adopted to comprehensively assess both quantitative metrics related to digital literacy levels and qualitative insights into the social experiences of individuals.

The study's significance lies in its potential to inform policies aimed at bridging the digital divide in Canada's rural areas. By investigating the relationship between digital literacy and social inclusion, the research aims to contribute practical recommendations for community-based programs and governmental initiatives that foster a digitally inclusive and socially connected rural Canada.

CHAPTER 11

Expected Outcomes of the Research of a Research Proposal

ABSTRACT
This chapter conducts an in-depth exploration of the pivotal concept of "expected outcomes of the research" within a research proposal. It delves into the significance of this section, which provides insight into the anticipated results and contributions of the study. Furthermore, the chapter navigates key considerations when formulating expected outcomes, underscoring their importance in aligning the study with its objectives. It elaborates on why including a dedicated section on expected outcomes enhances the proposal's transparency and demonstrates the researcher's strategic vision. Additionally, practical guidance is offered on crafting a robust "Expected Outcomes" section that is realistic, achievable, and aligned with the study's design. To exemplify these principles, a case study is presented, focusing on a research proposal investigating the impact of online education on student learning outcomes.

KEYWORDS: Expected outcomes, anticipated results, contributions, transparency, strategic vision, realistic, achievable, case study, online education, student learning outcomes.

11.1 What Do We Mean by the Expected Outcomes of the Research?

The "expected outcomes of the research" in a research proposal refer to the anticipated results, findings, or conclusions that the researcher envisions as a result of conducting the proposed study. It is a crucial section of the research proposal as it outlines the specific contributions and impacts that the research aims to achieve. By describing the expected outcomes, researchers can convey the potential value and significance of their study to reviewers, funders, and other stakeholders.

11.2 Key Points to Consider

Here are some key points to consider when describing the expected outcomes in a research proposal:

- Align with Research Objectives: The expected outcomes should directly relate to the research objectives stated in the proposal. Each research objective should have corresponding expected outcomes that demonstrate how the objective will be achieved.
- Be Specific and Measurable: Describe the expected outcomes in specific terms. Avoid vague or general statements and use measurable indicators whenever possible. This allows for a clear evaluation of the research's success.
- Link to Hypotheses or Research Questions: If the research proposal includes hypotheses or research questions, the expected outcomes should be linked to these. Explain how the research will answer the questions or test the hypotheses.
- Contribute to Knowledge: Emphasize how the expected outcomes will contribute to the existing body of knowledge in the field. Will the research confirm existing theories, challenge established concepts, or introduce new insights?
- Practical Applications: If applicable, highlight the potential practical applications of the research findings. Describe how the outcomes can be used to address real-world problems or inform policy decisions.
- Address Potential Impact: Discuss the potential impact of the research outcomes on various stakeholders, communities, or industries. This could include social, economic, environmental, or health-related impacts.
- Consider Multiple Scenarios: Acknowledge that research outcomes may vary, and there could be alternative or unexpected results. While outlining the expected outcomes, also mention any potential limitations or alternative scenarios.

Stay Realistic: Ensure that the expected outcomes are realistic and achievable within the scope and resources of the proposed study. Avoid making grand claims that cannot be substantiated.

Importance and Significance: Clearly articulate why the expected outcomes are important and how they address gaps or issues in the field of study. Explain the relevance and significance of the research findings.

Potential Follow-Up Research: If the research opens new avenues for further investigation, mention potential follow-up studies that can build upon the expected outcomes.

By providing a well-thought-out and compelling description of the expected outcomes, researchers can demonstrate the potential value and impact of their proposed study. Reviewers and funders often consider the expected outcomes as a critical factor in evaluating the feasibility and significance of the research proposal.

11.3 Why Will We Include Expected Outcomes?

Including "Expected outcomes" in a research proposal serves several important purposes and benefits the overall proposal in the following ways:

Clarity of Research Goals: By outlining the expected outcomes, researchers provide clarity on the specific goals and objectives of their study. This helps reviewers and stakeholders understand the purpose and focus of the research.

Justification of the Research: Expected outcomes demonstrate the anticipated value and significance of the proposed study. It helps justify why the research is worth conducting and what potential contributions it can make to the field.

Evaluation of Feasibility: Reviewers can assess the feasibility of the research based on the expected outcomes. Clear and achievable outcomes show that the research is well-planned and can be conducted within the proposed timeframe and resources.

Relevance to Funding Agencies: Funding agencies often look for research proposals that have tangible outcomes and practical applications. Including expected outcomes makes the proposal more appealing to potential funders.

Research Design and Methodology: Expected outcomes help in designing appropriate research methodologies. Researchers can choose methods and data collection techniques that align with the intended outcomes.

Guidance for Data Analysis: Expected outcomes provide direction for data analysis. Researchers can focus on analyzing data in ways that address the research objectives and yield relevant findings.

Potential Impact and Benefits: By discussing the potential impact of the research outcomes, researchers can demonstrate the benefits the study may bring to society, industry, or academia.

Transparency and Credibility: Clearly stated expected outcomes enhance the transparency of the research proposal. It shows that the researchers have thought through their study and are not making unfounded claims.

Alignment with Research Questions/Hypotheses: Expected outcomes are directly related to the research questions or hypotheses, establishing a logical connection between the research design and the research objectives.

Communication with Stakeholders: Including expected outcomes helps researchers effectively communicate the intended results to stakeholders, collaborators, and potential participants.

Potential for Dissemination and Publication: Clear expected outcomes increase the likelihood of successful dissemination and publication of research findings, as reviewers and journals are more interested in studies with meaningful outcomes.

Opportunity for Course Correction: Outlining expected outcomes allows researchers to recognize potential challenges or limitations in their study design early on, enabling them to make necessary adjustments.

In summary, including "Expected outcomes" in a research proposal is essential for communicating the goals, significance, and potential impact of the proposed study. It helps reviewers and stakeholders assess the relevance and feasibility of the research and strengthens the overall credibility of the research proposal.

11.4 How Can We Write a Good Section on the Expected Outcomes?

Writing a good section on "Expected Outcomes" of the research in a research proposal involves providing a clear, concise, and compelling description of the anticipated results, findings, and contributions of the study. Here are some steps and tips to help you craft an effective "Expected Outcomes" section:

- Link to Research Objectives: Begin the section by restating the research objectives briefly. Then, explain how each objective will lead to specific expected outcomes. Ensure that the expected outcomes directly align with the stated research objectives.
- Be Specific and Measurable: Clearly state the expected outcomes in specific terms. Use measurable indicators or metrics when possible to quantify the outcomes. This allows for a more precise evaluation of the research's success.
- Relate to Research Questions or Hypotheses: If your research proposal includes research questions or hypotheses, explain how the expected outcomes will provide answers to these questions or test the hypotheses.
- Contribution to Knowledge: Emphasize how the expected outcomes will contribute to the existing body of knowledge in the field.

Describe whether the research will confirm, refute, or extend existing theories or concepts.
- Practical Applications: Discuss the potential practical applications of the research findings. Explain how the outcomes can be used to address real-world problems, inform policy decisions, or improve practices in relevant domains.
- Addressing Gaps in the Literature: Highlight how the expected outcomes will address gaps or limitations in the existing literature. Demonstrate the relevance of the study within the broader academic context.
- Potential Impact: Discuss the potential impact of the research outcomes on various stakeholders, communities, industries, or fields of study. Consider social, economic, environmental, or health-related impacts.
- Feasibility and Realism: Ensure that the expected outcomes are realistic and achievable within the scope and resources of the proposed study. Avoid making unfounded claims or overly optimistic projections.
- Alternative Scenarios: Acknowledge that research outcomes may vary and that alternative or unexpected results are possible. Mention any potential limitations or alternative scenarios and how they will be handled.
- Link to Research Design and Methodology: Explain how the research design and methodology support the expected outcomes. Discuss how the chosen methods are suitable for achieving the intended results.
- Value Proposition: Clearly articulate why the expected outcomes are valuable and significant. Discuss the potential contributions of the research to the field and its broader implications.
- Importance of Dissemination: Emphasize the importance of disseminating the research outcomes. Discuss plans for sharing the findings with relevant audiences, including academic communities, policymakers, and the public.

Potential for Follow-Up Research: If the research opens new avenues for further investigation, mention potential follow-up studies that can build upon the expected outcomes.

Use Clear Language: Write clearly and straightforwardly, avoiding jargon or technical language that may be difficult for non-specialists to understand.

By following these steps and incorporating these tips, you can create a compelling "Expected Outcomes" section that effectively communicates the anticipated value and significance of your research. This section enhances the overall quality of the research proposal and helps reviewers and stakeholders appreciate the potential contributions of the study.

10.5 Example: Research Proposal on the Impact of Online Education on Student Learning Outcomes

Improved Academic Performance: The study is expected to reveal a positive correlation between online education and improved academic performance. We anticipate that students who engage in online learning will demonstrate higher grades and overall academic achievement compared to traditional classroom-based learners.

Enhanced Student Engagement: The research aims to identify factors that contribute to increased student engagement in online education. By understanding these factors, we anticipate that educators and institutions can design more interactive and immersive online learning experiences, leading to higher levels of student engagement.

Identification of Effective Pedagogical Strategies: The study will explore various pedagogical approaches used in online education. We expect to identify the most effective strategies that facilitate better student understanding, knowledge retention, and critical thinking skills.

- Impact on Student Motivation and Self-Regulation: The research will investigate the impact of online education on student motivation and self-regulation. We anticipate that the flexible nature of online learning may empower students to take more ownership of their learning, leading to increased self-regulation and greater motivation to succeed academically.
- Accessibility and Inclusivity: The study will examine how online education can improve access to education for diverse student populations, including individuals with disabilities or those in remote areas. We expect to find that online learning platforms can increase educational opportunities and promote inclusivity.
- Institutional Readiness and Faculty Training: The research will assess the readiness of educational institutions to implement online education effectively. We anticipate that the findings will highlight the importance of providing faculty with adequate training and support to ensure successful online course delivery.
- Challenges and Barriers: The study will identify challenges and barriers faced by both students and educators in the context of online education. By understanding these obstacles, we aim to provide recommendations to overcome them and enhance the overall quality of online learning experiences.
- Impact on Workforce Preparedness: The research will explore the extent to which online education prepares students for the workforce. We expect to find evidence of how online learning equips learners with relevant skills and competencies sought by employers.
- Quality Assurance and Accreditation: The study will investigate the measures taken to ensure the quality and accreditation of online education programs. We anticipate that the research will contribute to ongoing discussions on ensuring the credibility and recognition of online degrees and certifications.
- Policy Recommendations: Based on the research findings, the study will provide policy recommendations to educational policymakers and institutions on how to optimize online education delivery. These recommendations will aim to foster a conducive

Expected Outcomes of the Research Proposal

environment for online learning and encourage its integration into mainstream educational systems.

These expected outcomes provide a glimpse of the potential impacts and contributions of the proposed research on the topic of online education and student learning outcomes. They are specific, measurable, and relevant to the research objectives, demonstrating the significance of the study and its potential value to the field of education.

CHAPTER 12

Writing Gantt Chart and Budget of a Research Proposal

ABSTRACT
This chapter explores the dynamic aspects of crafting both a Gantt Chart and a Budget within a research proposal. It begins by elucidating the fundamental concepts, with a Gantt Chart serving as a visual timeline outlining the research's stages and a Budget detailing financial allocations for the study. Furthermore, the chapter navigates the essential components of a Gantt Chart, emphasizing its role in structuring and tracking research progress. The benefits of incorporating a Gantt Chart are highlighted, including improved project management and communication. Practical insights are offered into the art of constructing an effective Gantt Chart that is both comprehensive and visually intuitive. To illustrate these principles, a sample Gantt Chart is presented. Additionally, the chapter seamlessly transitions to explore the concept of a budget plan, clarifying its importance in managing resources and ensuring research feasibility. The major components of a budget are outlined, showcasing the areas of financial consideration. A budget example is provided to illustrate these concepts. Lastly, the qualities of a strong budget plan are discussed, encompassing factors such as accuracy, transparency, and adaptability, ensuring that the budget serves as a reliable guide throughout the research journey.

KEYWORDS: Gantt chart, budget, communication, feasibility, resources, qualities, accuracy, transparency, adaptability

12.1 What Is a Gantt Chart?

A Gantt chart in a research proposal is a visual representation of the project timeline, outlining the tasks, milestones, and activities that will be undertaken throughout the research project. It is a valuable project management tool that helps researchers and project teams plan, schedule, and track the progress of their research activities. The chart takes its name from its creator, Henry Gantt, an American engineer and management consultant who developed it in the early twentieth century. The Gantt chart is typically presented as a horizontal bar chart, where each bar represents a specific task or activity. The length of the bar corresponds to the duration of the task, and the chart is divided into time intervals, such as days, weeks, or months. The chart provides a visual overview of the project schedule, showing the start and end dates of each task, as well as the dependencies between tasks.

12.2 Key Components of a Gantt Chart

- Tasks and Activities: Each task or activity in the research project is represented by a bar on the Gantt chart. These tasks can include literature review, data collection, data analysis, writing the report, etc.
- Duration: The length of each task bar represents the estimated time it will take to complete the activity. The duration can be measured in days, weeks, or months, depending on the scale of the project.
- Start and End Dates: The Gantt chart shows the start and end dates for each task, indicating when the activity is planned to begin and when it is expected to be completed.
- Dependencies: Tasks may have dependencies, meaning that the completion of one task is dependent on the completion of another task. The Gantt chart visualizes these dependencies, helping to identify critical paths and potential bottlenecks in the project.
- Milestones: Milestones are significant events or achievements in the project timeline. They are represented by diamond-shaped symbols on the Gantt chart and help track major progress points in the research.
- Resource Allocation: Some Gantt charts also include information about resource allocation, indicating which team members or resources are responsible for each task.

12.3 Benefits of Using a Gantt Chart

Effective Planning: The Gantt chart helps researchers plan the project timeline, ensuring that tasks are organized in a logical sequence and that deadlines are achievable.

Visual Representation: The visual nature of the Gantt chart makes it easy to understand and communicate the project schedule to stakeholders and team members.

Tracking Progress: Researchers can track the progress of their research project by updating the Gantt chart regularly. It helps identify delays and allows for timely adjustments.

Resource Management: The Gantt chart enables researchers to allocate resources efficiently, ensuring that team members are assigned to tasks appropriately.

Time Management: By breaking down the research project into specific tasks and assigning timeframes, the Gantt chart facilitates effective time management.

Overall, the Gantt chart is a valuable tool that enhances the organization, coordination, and execution of a research project, making it an essential component of a research proposal.

12.4 How to Write an Effective Gantt Chart?

Writing an effective Gantt chart involves careful planning and attention to detail. Here are the steps to create an impactful Gantt chart for your research proposal:

Identify Tasks and Activities: Start by listing all the tasks and activities that need to be completed for your research project. Break down the project into smaller, manageable components. Tasks could include literature review, data collection, data analysis, writing chapters, and preparing presentations.

Set Realistic Durations: Estimate the time required to complete each task. Be realistic and consider any potential challenges or delays that may arise during the research process. Ensure that the durations are reasonable and achievable.

Determine Dependencies: Identify any dependencies between tasks. Some tasks may be sequential, meaning one task must be completed before the next can begin. Others may be parallel, allowing for simultaneous work.

Establish Milestones: Define important milestones in the project timeline. Milestones are significant achievements or completion points in the research. They could be the submission of specific chapters, data collection completion, or the presentation of findings.

Choose the Gantt Chart Format: Decide on the format and scale of the Gantt chart. You can create the chart using various tools, such as spreadsheet software (e.g. Microsoft Excel) or project management software. Choose a time scale that best suits your project's duration (e.g. days, weeks, and months).

Create the Gantt Chart: Plot the tasks and activities on the Gantt chart, arranging them in chronological order. Use horizontal bars to represent each task, and position them according to their start and end dates. Add milestone points as diamond-shaped symbols.

Use Color Codes or Labels: Consider using color codes or labels to differentiate different phases or types of tasks. This enhances clarity and makes the Gantt chart easier to understand.

Include Resource Allocation: If relevant, indicate the responsible team members or resources for each task. This helps with resource management and accountability.

Update and Maintain the Chart: As the research progresses, update the Gantt chart regularly to reflect the actual progress. Adjust task durations, add new tasks if necessary, and mark completed tasks and milestones.

Visualize Dependencies: Clearly show task dependencies using arrows or connecting lines between dependent tasks. This allows you to identify critical paths and potential delays.

- Keep it Simple and Clear: Avoid cluttering the Gantt chart with too much detail. Keep the presentation simple and easy to read. Use concise task descriptions and avoid overwhelming the chart with excessive information.
- Use Annotations: Consider using annotations or notes to provide additional information or explanations for specific tasks or milestones.

By following these steps, you can create an effective Gantt chart that serves as a valuable project management tool for your research proposal. A well-designed Gantt chart allows you to visualize the project timeline, track progress, and ensure that your research stays on schedule.

12.5 An Example of a Gantt Chart [Eight Months Project] with Work Schedule and Staffing Schedule

Deliverable/Month	1	2	3	4	5	6	7	8
Contact Sign and detail work plan development	■							
Literature review	■	■						
Inception report, methodology and data collection tools development and tools sharing with IEM Unit		■						
Recruitment of enumerators, training, and pilot study			■					
Data collection				■	■			
Data editing and data drafting						■		
Sharing major/key findings with IEM Unit						■		
Report writing							■	
Report Presentation (Dissemination workshop)							■	
Final Report Submission with the incorporation of feedback								■

Note: This is a tentative activity plan, and it could be changed upon any unavoidable circumstances. Accordingly, the remaining activities can be shifted.

A comprehensive and effective Gantt Chart requires the inclusion of two crucial components in the financial proposal: the work schedule and the staff schedule. These aspects are pivotal in ensuring the successful execution of the project, as they align the project's timeline and resource allocation with the overall objectives.

Work Schedule: The work schedule encompasses the breakdown of tasks, activities, and milestones that need to be accomplished throughout the project's lifecycle. Each task is assigned a specific start and end date, allowing for a clear visualization of the project's progress over time. This schedule aids in identifying potential bottlenecks, overlaps, or gaps in the project timeline. When integrated with the Gantt Chart, the work schedule provides a visual representation of task dependencies, helping project managers allocate resources effectively and ensure a smooth flow of work.

Staff Schedule: The staff schedule outlines the allocation of team members to various tasks and activities according to their skills, expertise, and availability. It ensures that the right people are assigned to the right tasks at the right time, optimizing resource utilization and preventing the potential overburdening of certain team members. Additionally, the staff schedule considers factors such as team members' roles, responsibilities, and capacity to manage their workload efficiently. By integrating the staff schedule into the Gantt Chart, project managers can easily monitor the distribution of human resources and make necessary adjustments if any imbalances arise.

In this Gantt chart, the research tasks are organized in chronological order, and each task is represented by a horizontal bar. The duration of each task is indicated in weeks, and the start date of each task is provided. The tasks in the Gantt chart represent different stages of the research project, including literature review, proposal development, ethical approval, participant recruitment, data collection, intervention implementation, data analysis, report writing, and presentation preparation. The chart also includes milestone points, such as obtaining ethical approval and finalizing the research report. This Gantt chart provides a visual overview of the project timeline and allows the research team to track progress, manage deadlines, and identify potential delays.

Furthermore, the financial proposal's integration with the Gantt Chart ensures that the project's budget aligns with the project's timeline and resource allocation. For instance, linking the staff schedule with the salary details of team members allows for accurate budget forecasting and cost estimation. This integration empowers project managers to monitor not only the progress of tasks but also the associated costs, ensuring that the project remains within budgetary constraints.

12.5.1 Work Schedule [Twelve Months Project]

Deliverable/Month	1 Col=1 month											
	1st	2nd	3rd	4th	5th	6th	7th	8th	9th	10th	11th	12th
Contact Sign and detail work plan development	■											
Literature review	■	■										
Inception report, methodology and data collection tools development and tools sharing with respective organization/ funding agency		■	■									
Recruitment of enumerators, training, and pilot study			■									
Data collection			■	■	■	■						
Data editing and data drafting					■	■						

Deliverable/Month	1 Col=1 month											
	1st	2nd	3rd	4th	5th	6th	7th	8th	9th	10th	11th	12th
Sharing major/ key findings with respective organization/ funding agency							■					
Report writing								■	■	■	■	
Report Presentation (Dissemination workshop)												■
Final Report Submission with the incorporation of feedback												■

12.5.2 Staffing Schedule

Sl. No.	Name of Staff		Staff-month input by month (1 Col=10 days)												Total staff-month input		
			1	2	3	4	5	6	7	8	9	10	11	12	Home	Field	Total
1	Name: .... Team Leader	Home	■	■	■	■	■	■	■	■	■	■	■	■	3.75		4.0
		Field							■							0.25	
2	Name: .... Team Member	Home			■	■	■	■	■	■	■	■	■	■	3.25		4.0
		Field								■						0.25	
3	Name: .... Team Member	Home	■	■	■	■	■	■	■	■	■	■	■		3.25		4.0
		Field							■	■						0.25	
4	Name: .... Team Member	Home	■	■	■	■	■	■	■	■	■	■	■	■	3.25		4.0
		Field						■	■	■						0.25	
5	Name: .... Team Member	Home	■	■	■	■	■	■	■	■	■	■	■	■	3.25		4.0
		Field							■	■						0.25	
6	Statistician, computer compose & data analysis	Home									■	■	■	■	4.0		4.0
		Field							■	■							
7	Enumerators (24)	Home														24.0	24.0
		Field						■	■	■	■						
8	Supervisors (4)	Home														4.0	4.0
		Field						■	■	■	■						
	Total																52.0

12.6 Definition of Budget of a Research Proposal

A budget in a research proposal is a detailed financial plan that outlines the estimated costs and expenses associated with conducting the proposed research study. It includes anticipated expenditures for various aspects of the research, such as personnel salaries, equipment, materials, participant incentives, travel, data analysis software, and other relevant costs. The budget serves to demonstrate the feasibility of the research by showing how the requested funds will be allocated and managed to accomplish the study's goals. In some contexts, the budget plan within a research proposal is often referred to as a "financial proposal." This term emphasizes the financial aspect of the proposal and highlights the detailed breakdown of costs and expenses required to conduct the research. Just like the budget plan, the financial proposal provides a comprehensive overview of how the funds will be allocated to support the research activities, ensuring transparency and accountability in the use of resources.

12.7 Why a Budget Plan Is Important?

A budget plan holds significant importance within a research proposal for several reasons:

- Feasibility Assessment: The budget plan allows an evaluation of the practicality of the research endeavor by determining whether the available funds are sufficient for successful project completion.
- Resource Allocation: By delineating how finances will be distributed across various research aspects such as personnel, equipment, travel, and participant compensation, the budget plan ensures equitable allocation of resources.
- Transparency and Accountability: An in-depth budget plan fosters transparency by providing a clear breakdown of how funds will be

utilized. This transparency enhances accountability in handling research finances.

Realism: A comprehensive budget plan showcases meticulous consideration of the research's financial demands, highlighting the researchers' grounded comprehension of project costs.

Credibility: An articulated budget plan bolsters the proposal's credibility, reassuring reviewers, sponsors, and stakeholders of the researchers' competence in managing financial components.

Guidance: Serving as a financial roadmap, the budget plan aids researchers in adhering to financial limitations, thus facilitating successful research execution.

Competitive Proposals: Well-structured budget plans enhance proposal competitiveness, as funders are more inclined to support projects with logically justified and well-managed financial needs.

Negotiation and Approval: Post-approval, the budget plan becomes foundational for securing the required funding, providing a baseline for negotiation and financial discussions.

In essence, the budget plan ensures financial viability, transparency, and effective management of resources, bolstering the overall quality of the research proposal and its potential for successful implementation.

12.8 What Are the Major Components or Contents of a Budget?

The major components or contents of a budget in a research proposal typically include:

Personnel Costs: This includes salaries or stipends for researchers, assistants, technicians, and administrative staff directly involved in the research.

- Equipment: Costs related to purchasing, leasing, or maintaining specialized equipment required for the research.
- Materials and Supplies: Expenses for consumable items like chemicals, reagents, laboratory supplies, paper, pens, and other necessary materials.
- Travel: Funds allocated for research-related travel, such as attending conferences, fieldwork, data collection, and collaborations.
- Participant Compensation: If the research involves human participants, compensation or incentives may be allocated for their time and participation.
- Data Collection and Analysis: Costs associated with data collection tools, software licenses, data processing, and statistical analysis.
- Publication and Dissemination: Funding for presenting findings at conferences, publishing research papers, and producing any dissemination materials.
- Consultation and Services: Fees for external consultants, expert advice, or specialized services required for the research.
- Facility or Space Rental: If research requires renting specific facilities or space, these costs should be included.
- Overhead or Indirect Costs: Institutional costs not directly tied to the project, often calculated as a percentage of direct costs.
- Contingency Fund: A reserve for unforeseen expenses or changes in the research plan.
- Institutional Review Board (IRB) Fees: If human subjects are involved, there might be fees associated with ethical review processes.
- Miscellaneous: Any other relevant expenses that don't fall neatly into the above categories.

Each component should be clearly described in the budget with its associated costs, quantities, and justifications. Moreover, it's crucial to provide detailed explanations for each expense to demonstrate that the proposed budget is well-considered and necessary for the successful execution of the research project. Always adhere to the guidelines and formatting requirements provided by the funding agency or institution.

12.9 An Example of a Budget for a Research Proposal

12.9.1 Summary of Costs

Cost Component	Costs
Staff Remuneration[a]	17,30,000
Reimbursable Expenses[1]	12,50,000
Total	29,80,000 (Twenty-nine lac and eighty thousand)

[a] Staff Remuneration, Reimbursable Expenses, and Taxes must coincide with relevant Total Costs

12.9.2 Breakdown of Staff Remuneration

[Information to be provided in this form shall be used to establish Payments to the Consultants by the Client]

Name[a]	Position[b]	Staff-month Rate[c]	Input[c] (Staff-months)	[Indicate Sub Cost for each staff][d]
Staff				
Name:...............	Team Leader		160,000	3,20,000
Name:...............	Team Member		80,000	1,60,000
Name:...............	Team Member	Head office	80,000	1,60,000
Name:...............	Team Member		80,000	1,60,000
Name:...............	Team Member		80,000	1,60,000

Name[a]	Position[b]	Staff-month Rate[c]	Input[c] (Staff-months)	[Indicate Sub Cost for each staff][d]
		Field office		
–	Supervisor (4)		30,000	1,20,000
	Numerators (24)		25,000	6,00000
	Statistician, computer composer, and data analysis		50,000	50,000
			Total Costs	17,30,000

[We hereby confirm that we have agreed to pay the Staff Members listed, who will be involved in this assignment, the remuneration and away from Head office Allowances (if applicable) as indicated above].

[a] Professional Staff should be indicated individually; Support Staff should be indicated per category (e.g. draftsmen and clerical staff).

[b] Positions must coincide with the ones

[c] Indicate the total expected input of staff and staff-month rate required for carrying out the activity indicated in the Form.

[d] For each staff indicates the remuneration. Remuneration = Staff-month Rate x Input.

12.9.3 *Breakdown of Reimbursable Expenses*

[Information to be provided in this form shall be used to establish Payments to the Consultants by the Client]

Writing Gantt Chart and Budget of a Research Proposal

N°	Description[a]	Unit	Unit Cost[b]	Quantity	[Indicate sub-cost for each item][c]		
	Per diem allowances (TL & members)	Day	3,000 (5 days each)	5	75,000		
	Travel expenses	Trip	–	–	1,00,000		
	Communication costs between [Dhaka to the sampled area] and [Sampled areas to Dhaka]	–	–	–	35,000		
	FGDs and KIIs				1,00,000		
	Drafting, reproduction of reports				5,00,000		
	Equipment, instruments, etc. (Computer)				50,000		
	Workshop for feedback on a draft report				2,00,000		
	Use of computers, software				50,000		
	Other transportation costs				20,000		
	Office rent, clerical assistance				1,00000		
	Others (specify)				20,000		
Total Costs					12,50,000		

[a] Delete items that are not applicable or add other items.
[b] Indicate unit cost.
[c] Indicate the cost of each reimbursable item. Cost = Unit Cost × Quantity.
[d] Usually Physical Contingency shall not exceed five percent (5%)

12.10 Qualities or Characteristics of a Good Budget Plan

Creating a good budget plan in a research proposal involves several qualities and characteristics that contribute to its effectiveness and credibility. Here are some key qualities:

- Accuracy and Realism: Ensure that the budget reflects accurate cost estimates based on research needs. Overestimating or underestimating costs can lead to credibility issues and hinder the successful execution of the project.
- Thoroughness: Include all relevant expenses, even if they seem minor. A comprehensive budget plan shows that you've considered various aspects of the research and have a clear understanding of its financial requirements.
- Justification: Provide clear and concise explanations for each budget item. Justify why each expense is necessary for the research and how it directly contributes to achieving the project's goals.
- Transparency: Make the budget transparent by breaking down costs and providing itemized details. Transparency builds trust and helps reviewers or funders understand how funds will be allocated.
- Consistency: Ensure that the budget aligns with the project's objectives and methodology. Inconsistencies between the research plan and the budget can raise concerns about the proposal's coherence.
- Funding Guidelines: Adhere to any specific guidelines or requirements provided by the funding agency or institution. This includes formatting, allowable expenses, and any prescribed budget categories.
- Contingency: Include a contingency fund to account for unexpected expenses or changes in the research plan. A contingency demonstrates foresight and flexibility in handling unforeseen challenges.
- Reasonable Estimates: Use reasonable cost estimates based on market rates and industry standards. Unrealistically low estimates can raise questions about the feasibility of the research.

- Collaboration: If the research involves multiple collaborators or institutions, coordinate with them to create a unified and accurate budget that considers all parties' contributions.
- Precision: Be precise in calculations, ensuring that figures are accurate and properly calculated. Mathematical errors can undermine the credibility of your budget plan.
- Prioritization: If there are budget constraints, prioritize essential expenses that directly impact the research's quality and outcomes. Justify why certain items are crucial and can't be compromised.
- Review and Proofreading: Carefully review the budget plan for errors, inconsistencies, and omissions. A well-proofed budget plan indicates professionalism and attention to detail.
- Realistic Timeline: Align the budget with the project's timeline. Ensure that expenses are appropriately allocated across the research stages and that funds are available when needed.
- Clear Presentation: Present the budget in a clear and organized manner. Use tables, charts, and headings to enhance readability and comprehension.

By incorporating these qualities into your budget plan, you'll enhance the overall quality of your research proposal and increase the likelihood of securing funding for your project.

CHAPTER 13

References of a Research Proposal

ABSTRACT
This chapter serves as a comprehensive guide to managing references within a research proposal. It addresses the ideal quantity of references, considering both sufficiency and relevance to the research. The types of referencing are explored, encompassing in-text references and end referencing. Furthermore, the chapter navigates major reference styles or formats, showcasing examples of each to aid researchers in selecting an appropriate style for their proposal. By offering practical insights into effectively organizing and presenting references, the chapter ensures that the research proposal is grounded in scholarly discourse and maintains a high standard of academic integrity.

KEYWORDS: References, quantity, sufficiency, relevance, in-text references, end referencing, reference styles, scholarly discourse, academic integrity

13.1 How Many References Are Ideal for a Research Proposal?

The ideal number of references for a research proposal can vary depending on several factors, including the scope of the research, the depth of the literature review, and the specific requirements of the funding agency or institution to which the proposal is submitted. There is no fixed rule for the exact number of references in a research proposal, but some general guidelines can be followed.

>Thorough Literature Review: A research proposal should include a thorough literature review to demonstrate the researcher's understanding of the existing body of knowledge related to the research topic. The literature review should be comprehensive and provide relevant sources to support the research rationale and objectives.

- Balance Between Quality and Quantity: The focus should be on the quality of the references rather than the quantity. Including many references without relevant or reputable sources may not enhance the proposal's credibility. Instead, aim to cite key studies and seminal papers that directly relate to your research.
- Target Audience: Consider the target audience of the research proposal. If it is intended for an academic committee or research funding agency, a more extensive literature review with a higher number of references may be expected. However, for some other proposals, a concise yet impactful review may be sufficient.
- Recent and Relevant Sources: Include recent publications that reflect the current state of research in your field. However, do not overlook relevant older sources that have had a significant impact on the topic.
- Cite Primary Sources: Whenever possible, cite primary sources (original research papers) rather than relying solely on secondary sources (review articles or textbooks).
- Consistency and Cohesion: Ensure that the references cited are relevant to the research objectives and contribute to a cohesive narrative in the proposal.
- Follow Guidelines: If you are submitting the research proposal to a specific funding agency or institution, review their guidelines or requirements for the number of references they expect.

As a rough guideline, a research proposal for a moderate-sized project could have anywhere from fifteen to fifty references. However, the actual number will depend on the research field, the complexity of the study, and the depth of the literature review required to justify the research objectives.

It is essential to strike the right balance between providing enough references to support your research and avoiding an excessive number that may make the proposal overly dense or challenging to read. Focus on presenting a compelling and evidence-based rationale for your research while adhering to the specific requirements and expectations of your target audience.

13.2 Types of Referencing: Text References and End Referencing

The two main types of referencing commonly used in academic writing are "in-text references" (also known as "parenthetical references" or "citation within the text") and "end referencing" (also known as "bibliographic referencing" or "citation at the end"). Both types serve the purpose of providing proper credit to the sources of information cited in the research paper or academic document.

13.2.1 In-text References (Parenthetical References/Citations within the Text)

In-text references are citations placed within the body of the text, typically within parentheses or brackets. They are used to indicate the source of specific information, ideas, or quotes presented in the text. In-text references usually include the author's last name, the publication year, and sometimes the page number (for direct quotes). Different academic disciplines may have specific formatting styles for in-text references, such as APA (American Psychological Association), MLA (Modern Language Association), or Chicago/Turabian style.

Example (APA Style):
According to Smith (2020), "research proposals play a crucial role in the research process" (p. 15).

13.2.2 End Referencing (Bibliographic Referencing/Citation at the End)

End referencing involves providing a comprehensive list of all the sources cited in the research paper or document at the end of the work, usually in a separate section titled "References," "Bibliography," or "Works Cited." End referencing provides detailed information about each source, including the author's name, publication year, title of the work, publisher, and other relevant publication details.

Example (APA Style):

References:

Smith, J. (2020). *Writing a Research Proposal: A Comprehensive Guide.* Academic Press.

End referencing serves as a complete list of sources, allowing readers to locate the original works if they want to explore the cited sources further.

Both in-text references and end referencing are essential components of academic writing as they ensure proper acknowledgment of the sources used and avoid plagiarism. The choice between using in-text references and end referencing depends on the citation style required by the academic institution or the preference of the author. Different citation styles have their specific guidelines for formatting in-text references and end referencing, so it's crucial to follow the appropriate style consistently throughout the document.

13.3 Major Reference Styles or Format with Some Examples

There are several major reference styles or formats used in academic writing, each with specific guidelines for citing sources. Some of the most common reference styles include:

13.3.1 *APA (American Psychological Association) Style*

APA style is widely used in the social sciences, psychology, education, and other fields. In-text references in APA style typically include the author's last name and the publication year in parentheses. The end referencing is provided in a list of "References" at the end of the document, providing detailed publication information.

Example (In-text): According to Smith (2020), "research proposals play a crucial role in the research process."

Example (End Referencing):
References:
Smith, J. (2020). *Writing a research proposal: A comprehensive guide.* Academic Press.

13.3.2 MLA (Modern Language Association) Style

MLA style is commonly used in the humanities, such as literature, arts, and language studies. In-text references in MLA style typically include the author's last name and the page number (if available) in parentheses. The end referencing is provided in a list of "Works Cited."

Example (In-text): According to Smith, "research proposals play a crucial role in the research process" (15).
Example (End Referencing):
Works Cited:
Smith, John. *Writing a Research Proposal: A Comprehensive Guide.* Academic Press, 2020.

13.3.3 Chicago Manual of Style (CMS)/Turabian Style

Chicago/Turabian style is used in various disciplines, including history, literature, and social sciences. It has two documentation systems: the Notes and Bibliography system (used in humanities) and the Author-Date system (used in social sciences). In-text references in the Notes and Bibliography system consist of superscript numbers in the text, linked to footnotes or endnotes, while the Author-Date system uses parenthetical in-text citations with the author's last name and publication year.

13.3.4 Example (Notes and Bibliography System)

According to Smith, research proposals play a crucial role in the research process.[1]

Example (End Referencing):
Bibliography:
Smith, John. *Writing a Research Proposal: A Comprehensive Guide.* Academic Press, 2020.

13.3.5 IEEE (Institute of Electrical and Electronics Engineers) Style

IEEE style is commonly used in engineering, computer science, and related fields. In-text references in IEEE style use bracketed numbers in the text, corresponding to the full citation in the reference list.

Example (In-text): According to Smith [1], "research proposals play a crucial role in the research process."
Example (End Referencing):
References:
[1] J. Smith, *Writing a Research Proposal: A Comprehensive Guide.* Academic Press, 2020.

13.3.6 Harvard Referencing Style

The Harvard style is used in various disciplines and is known for its author-date in-text citation format. In-text references in Harvard style include the author's last name and the publication year in parentheses.

Example (In-text): According to Smith (2020), "research proposals play a crucial role in the research process."
Example (End Referencing):
Smith, J. (2020). *Writing a Research Proposal: A Comprehensive Guide.* Academic Press.

Each reference style has its specific rules for citing different types of sources (books, journal articles, websites, etc.), so it's important to consult the official style guides for accurate and consistent referencing. Properly following a specific reference style ensures that the sources are cited accurately, the work is properly attributed, and academic integrity is maintained.

CHAPTER 14

Evaluation of a Research Proposal

ABSTRACT
This chapter delves into the critical process of evaluating research proposals, a pivotal stage in determining the viability and potential impact of a research project. Beginning with an exploration of essential evaluation criteria and the development of comprehensive frameworks, the chapter provides insights into the multifaceted aspects considered during the evaluation process. Moreover, it sheds light on the crucial role of peer review, emphasizing its significance in ensuring rigorous and unbiased assessments. Additionally, the chapter navigates the realms of both quantitative and qualitative assessment, addressing metrics such as budgetary considerations, timeline feasibility, and the robustness of proposed methodologies. Furthermore, by examining best practices in peer review and emphasizing the integration of quantitative and qualitative dimensions, this chapter equips researchers, reviewers, and evaluators with the tools to critically assess and enhance the quality of research proposals.

KEYWORDS: Research proposal evaluation, peer review, evaluation criteria, quantitative assessment, qualitative dimensions

14.1 Key Components of Evaluation Criteria

Evaluation criteria play a pivotal role in the rigorous assessment of research proposals, acting as the guiding framework against which the merits and shortcomings of a proposal are measured. This systematic and standardized approach ensures a fair and comprehensive evaluation process, providing valuable insights into the viability and potential impact of the proposed research.

Relevance to Research Objectives: In the evaluation of research proposals, a primary consideration is the alignment with defined research objectives. Evaluators meticulously examine the extent to which the proposed research addresses the core questions and objectives laid out in the proposal. A clear and direct connection between the identified research problem and the proposed methodology enhances the overall quality of the proposal.

Methodological Rigor: Another crucial aspect of the evaluation process is the assessment of methodological rigor. This involves a thorough scrutiny of the proposed research methodology to ensure its reliability and validity. Evaluators delve into the intricacies of the research design, data collection methods, and statistical or analytical approaches, aiming to ascertain the robustness of the proposed methodology and the credibility of potential research outcomes.

Feasibility and Timeliness: Evaluators carefully examine the feasibility of implementing the proposed research within the provided timeline and available resources. This component addresses the practical aspects of the research plan, evaluating whether the expectations set by the researcher are realistic and achievable. A well-defined timeline contributes to the overall assessment, reflecting the researcher's understanding of project management and resource utilization.

Innovation and Contribution to Knowledge: The evaluation process considers the degree of innovation inherent in the proposed research. Evaluators focus on how the research contributes to existing knowledge within the field. Attention is given to whether the proposal introduces novel perspectives, methodologies, or insights. This component highlights the significance of research in pushing the boundaries of knowledge and fostering intellectual advancements.

Clarity and Coherence: The clarity of a research proposal's structure and the coherence of its components are pivotal to the evaluation process. Evaluators assess how well the proposal is organized and articulated. A well-structured and clear proposal enhances the evaluator's ability to comprehend and assess the research plan effectively. This criterion emphasizes the importance of effective communication in research proposal writing.

Ethical Considerations: The ethical integrity of the proposed research is a critical aspect of the evaluation. Evaluators meticulously examine the proposal's adherence to ethical standards and guidelines. This includes considerations of participant consent, data privacy, and the overall ethical conduct of the research. The ethical component ensures that the proposed research respects the rights and well-being of participants and upholds the principles of research integrity.

These key components in the evaluation criteria provide a structured and comprehensive framework for assessing research proposals. The systematic analysis of relevance, methodological rigor, feasibility, innovation, clarity, and ethical considerations ensures a holistic evaluation that goes beyond mere academic scrutiny, contributing to the advancement of knowledge and the promotion of ethical research practices.

14.2 Development of Evaluation Frameworks

In the intricate process of evaluating research proposals, the development of robust evaluation frameworks is a fundamental aspect. These frameworks provide the structure and guidelines necessary for a systematic and comprehensive assessment of proposals, ensuring consistency and fairness in the evaluation process.

14.2.1 Key Components of Evaluation Frameworks

Criterion Definition:

- Clarification of Evaluation Criteria: The evaluation framework begins with a precise definition of the criteria that will be used to assess the research proposals. Each criterion is articulated in a way that aligns with the specific goals and expectations of the research.

The weighting of Criteria:

- Assigning Relative Importance: The development of an evaluation framework involves assigning weights to each criterion based on its relative importance. This weighting reflects the significance of each criterion in the overall assessment, allowing for a nuanced evaluation that considers the varying impacts of different components.

Scoring System:

- Establishing a Scoring Mechanism: An effective evaluation framework includes a scoring system that translates qualitative assessments into quantitative scores. This scoring mechanism provides a standardized method for evaluators to assign numerical values to different aspects of the proposal.

Rubric Creation:

- Detailed Assessment Guidelines: A well-crafted evaluation framework includes a rubric that provides detailed assessment guidelines for each criterion. The rubric breaks down the evaluation process, offering evaluators clear benchmarks and descriptors for different levels of performance.

Calibration of Evaluators:

- Ensuring Consistency: The development of an evaluation framework involves a calibration process to ensure consistency among evaluators. This may include training sessions, calibration exercises, and ongoing communication to align evaluators' interpretations of the criteria.

14.2.2 Importance of Evaluation Frameworks

Fairness and Consistency:

- Ensuring Fair Evaluation: Evaluation frameworks contribute to fairness and consistency in the assessment process. By clearly defining criteria, assigning weights, and establishing a scoring system, frameworks help mitigate subjectivity, ensuring that all proposals are evaluated against the same set of standards.

Transparency in Evaluation:

- Enhancing Transparency: Frameworks enhance the transparency of the evaluation process. They provide a clear roadmap for evaluators and proposal writers, making explicit the expectations and considerations that will guide the assessment.

Effective Communication:

- Facilitating Communication: The use of well-defined evaluation frameworks facilitates effective communication among evaluators, researchers, and stakeholders. A common language and understanding of evaluation criteria contribute to meaningful discussions and feedback.

Quality Assurance:

- Ensuring Quality Standards: Evaluation frameworks serve as tools for quality assurance. They set a benchmark for the expected quality of research proposals, guiding researchers in crafting proposals that align with established standards.

Continuous Improvement:

- Supporting Iterative Processes: The development of evaluation frameworks is not a static process. It supports iterative improvements by allowing for feedback and adjustments based on the evolving landscape of research practices and priorities.

14.3 Peer Review Process

Peer review stands as a cornerstone in the evaluation of research proposals, embodying a rigorous and collaborative mechanism for ensuring the quality, validity, and relevance of proposed research. This section explores the intricacies of the peer review process, shedding light on its significance, best practices, and the role it plays in shaping scholarly discourse.

14.3.1 Key Components of the Peer Review Process

Selection of Peers:

- Expertise Match: The peer review process begins with the careful selection of peers—individuals with expertise relevant to the subject matter of the research proposal. Matching the expertise of reviewers to the proposal's content ensures a thorough and informed assessment.

Anonymous or Open Review:

- Choice of Review Format: Peer review can be conducted anonymously (double-blind) or in an open format. Anonymity minimizes bias, while open review encourages transparency. The choice often depends on the preferences of the journal, funding agency, or institution overseeing the peer review.

Evaluation Criteria:

- Application of Defined Criteria: Reviewers apply predefined evaluation criteria, often outlined in the evaluation framework. These criteria typically encompass relevance to objectives, methodological rigor, ethical considerations, and other relevant aspects.

Constructive Feedback:

- Detailed and Constructive Comments: Reviewers provide detailed feedback, highlighting the strengths and weaknesses of the proposal. Constructive comments guide the researcher in refining and improving the proposal for eventual publication or funding.

Evaluation of a Research Proposal

Recommendations:

- Decision and Recommendations: Reviewers may recommend acceptance, revisions, or rejection of the proposal. Their recommendations play a pivotal role in the decision-making process by editors, funding agencies, or review committees.

14.3.2 Significance of the Peer Review Process

Quality Assurance:

- Ensuring Rigor and Quality: Peer review serves as a quality control mechanism, ensuring that only high-quality and credible research proposals advance to the next stage. It upholds the standards of academic and scientific rigor.

Validation of Research:

- Validation through Expert Scrutiny: The process of having peers scrutinize a proposal validates its scholarly merit. The endorsement of experts in the field enhances the credibility and reliability of the proposed research.

Improvement and Refinement:

- Iterative Process of Improvement: Peer review provides researchers with invaluable feedback for improving and refining their work. The iterative nature of this process contributes to the continuous enhancement of research proposals.

Gatekeeping Role:

- Guardians of Research Integrity: Peer reviewers act as guardians of research integrity, ensuring that proposed studies adhere to ethical standards, methodological rigor, and the broader ethical norms of the scientific community.

14.3.3 Challenges and Considerations

Bias and Fairness:

- Addressing Bias in Review: Efforts are made to address potential biases in the peer review process, including gender bias, cultural bias, or biases related to research methodologies. Transparency and diversity in the selection of reviewers contribute to mitigating these challenges.

Workload and Time Constraints:

- Managing Reviewer Workload: Reviewers often face heavy workloads, and time constraints can impact the thoroughness of the review. Balancing the demands on reviewers' time is a perpetual challenge in the peer review process.

Consistency in Reviews:

- Ensuring Consistency: Maintaining consistency among reviewers in their assessments is an ongoing consideration. Training, clear guidelines, and periodic calibrations help in aligning reviewers' judgments.

14.4 Best Practices in Peer Review

As the bedrock of scholarly evaluation, peer review is a dynamic process that benefits from a set of best practices designed to ensure fairness, transparency, and the delivery of constructive feedback. This section delves

Evaluation of a Research Proposal 201

into the key best practices that underpin an effective peer review process, acknowledging its role as a cornerstone in maintaining the integrity and quality of academic research.

14.4.1 Key Best Practices

Expertise Matching:

- Aligning Reviewers with Expertise: Assigning reviewers who possess expertise in the specific field of the research proposal is foundational. This ensures that evaluations are informed by a deep understanding of the subject matter.

Constructive Critique:

- Providing Constructive Feedback: Reviewers focus on delivering constructive criticism. Instead of merely identifying flaws, they offer insights that guide the researcher in improving the proposal. This approach fosters a collaborative and developmental aspect of the review process.

Timely Responses:

- Adherence to Timelines: Best practices emphasize the importance of timely responses. Both reviewers and the editorial or decision-making body commit to respecting deadlines, ensuring a streamlined and efficient review process.

Confidentiality and Integrity:

- Maintaining Confidentiality: The peer review process relies on confidentiality to encourage open and honest assessments. Reviewers commit to upholding the integrity of the review process by treating proposals and their content with confidentiality.

Ethical Considerations:

- Addressing Ethical Concerns: Reviewers are attuned to ethical considerations. This includes identifying potential ethical issues in the proposed research and ensuring that the researcher has addressed them appropriately.

Transparency and Openness:

- Transparent Communication: Transparent communication between reviewers, researchers, and editors is paramount. Clear articulation of evaluation criteria, expectations, and decisions fosters a transparent and accountable peer review process.

Diversity and Inclusion:

- Promoting Diversity in Reviewers: Ensuring diversity in the pool of reviewers contributes to a richer evaluation process. Diverse perspectives help mitigate biases and enhance the comprehensiveness of the review.

Continuous Training:

- Training and Professional Development: Reviewers are encouraged to engage in continuous training and professional development. This includes staying abreast of emerging methodologies, ethical considerations, and best practices in their respective fields.

14.4.2 Significance of Best Practices

Quality Enhancement:

- Improving the Quality of Reviews: Best practices aim to elevate the quality of reviews, moving beyond mere identification of issues to offering substantive suggestions for improvement. This enhances the overall quality of the research proposal.

Trust and Credibility:

- Building Trust in the Process: Adherence to best practices builds trust in the peer review process. Researchers, institutions, and the broader academic community rely on peer review to maintain credibility and integrity in scholarly publishing and funding.

Researcher Development:

- Supporting Researcher Development: Best practices are designed to support the development of researchers. Constructive feedback, guidance, and mentorship provided through peer review contribute to the continuous improvement of researchers' skills and methodologies.

Global Collaboration:

- Facilitating Global Collaboration: Best practices contribute to a globally collaborative research environment. By ensuring fairness, transparency, and inclusivity, the peer review process becomes a collaborative effort that transcends geographic boundaries.

14.4.3 Challenges and Considerations

Reviewer Workload:

- Managing Reviewer Workload: The challenge of managing the reviewer workload is a consideration. Implementing best practices involves addressing this challenge through strategies such as acknowledging and appreciating reviewers' contributions.

Balancing Consistency and Diversity:

- Balancing Consistency and Diversity: Achieving a balance between the consistency of reviews and the diversity of perspectives is an ongoing consideration. Best practices aim to integrate diverse viewpoints while maintaining a consistent standard of evaluation.

Adaptability to Emerging Trends:

- Adapting to Emerging Trends: Best practices need to be adaptable to emerging trends in research and scholarly communication. This requires a commitment to continuous improvement and flexibility in response to changing research landscapes.

14.5 Quantitative Metrics for Proposal Evaluation

Quantitative metrics provide a structured and objective approach to evaluating research proposals. In this chapter, we explore the utilization of quantitative measures in the assessment of proposals, emphasizing their significance, challenges, and the nuanced considerations required for a comprehensive evaluation.

14.5.1 Key Quantitative Metrics

Bibliometric Analysis:

- Citation Impact: Evaluating the citation impact of a researcher's previous work provides quantitative insights into the influence and visibility of their contributions. This metric is often used to gauge the researcher's standing in the academic community.

Publication Productivity:

- Number of Publications: The quantity of publications, especially in reputable journals, is a quantitative metric that reflects a researcher's productivity. This metric can indicate the extent to which a researcher has been actively contributing to their field.

Research Funding History:

- Amount of Research Funding Obtained: Assessing the amount of research funding a researcher has secured provides a quantitative measure of their success in attracting financial support. This metric can indicate the perceived value of their research by funding agencies.

Collaboration Networks:

- Collaboration Metrics: Quantitative measures of collaboration, such as co-authorship networks and collaboration indices, can provide insights into the researcher's engagement with the broader academic community. This includes the diversity and strength of their collaborative networks.

Research Impact Factor:

- Journal Impact Factor of Publications: The impact factor of journals where a researcher publishes is a quantitative indicator of the perceived influence of those publications. It reflects the average number of citations articles in a journal receive over a specific period.

14.6 Qualitative Assessment Components

In the qualitative assessment of research proposals, evaluators delve into the nuanced aspects that may not be easily quantifiable but are essential for understanding the depth, significance, and feasibility of the proposed research. This chapter explores the key qualitative assessment components commonly considered during the evaluation of research proposals.
Clarity and Coherence:

- Clear Expression of Ideas: Evaluators assess how well the proposal articulates its ideas. Clarity in expressing the research questions, objectives, and overall plan is crucial. A well-organized and coherent structure enhances the proposal's qualitative strength.

Conceptual Framework:

- Theoretical Foundation: Evaluators look for a robust conceptual framework that underpins the research. This involves assessing how well the proposal positions itself within existing theoretical perspectives and frameworks relevant to the research problem.

Research Design and Methodology:

- Comprehensive Methodology: Qualitative assessment of the research design involves evaluating the depth and appropriateness of the chosen methodology. Efficacy in addressing the research questions and the suitability of data collection and analysis methods are key considerations.

Critical Literature Review:

- Depth of Literature Review: The qualitative assessment includes an evaluation of the literature review's critical analysis. A strong literature review not only identifies relevant sources but also critically examines and synthesizes existing knowledge.

Intellectual Merit:

- Originality and Intellectual Rigor: Evaluators gauge the intellectual merit of the proposal, considering its originality and the intellectual rigor demonstrated. Proposals that contribute innovative ideas and exhibit a high level of intellectual engagement receive qualitative commendation.

Feasibility and Realism:

Evaluation of a Research Proposal

- Practicality of Implementation: Qualitative assessment involves an evaluation of the feasibility and realism of the proposed research. This includes considering the practical challenges and constraints that may be encountered during the research process.

Ethical Considerations:

- Adherence to Ethical Standards: The qualitative assessment includes a thorough examination of the proposal's ethical considerations. Evaluators look for evidence that the researcher has considered and addressed ethical issues related to participants, data, and the overall research process.

Potential for Impact:

- Contribution to Knowledge: Evaluators qualitatively assess the potential impact of the proposed research on the broader academic community and society. They consider how well the proposal articulates its contribution to knowledge and the potential influence it may have.

Researcher Qualifications:

- Researcher Expertise: The qualifications and expertise of the researchers involved in the project are qualitatively evaluated. Evaluator attention is given to the researchers' track record, relevant experience, and demonstrated capability to carry out the proposed research.

Communication Skills:

- Effective Communication: Qualitative assessment includes an evaluation of the researcher's communication skills. This involves assessing the clarity, persuasiveness, and overall effectiveness of the written proposal in conveying the research ideas.

Collaboration and Interdisciplinarity:

- Collaborative Elements: Qualitative assessment considers the potential for collaboration and interdisciplinarity. Proposals that demonstrate an openness to collaboration and interdisciplinary approaches receive positive qualitative recognition.

Adherence to Guidelines:

- Conformance to Guidelines: The extent to which the proposal adheres to the guidelines provided by the funding agency or institution is qualitatively assessed. A proposal that aligns with the specified format and requirements is viewed positively.

Overall Impression:

- Holistic Evaluation: Qualitative assessment involves forming an overall impression of the proposal. This holistic evaluation considers how well the various components of the proposal come together to form a compelling and well-reasoned research plan.

14.7 Evaluating Research Proposals Using a Point System

The use of a point system for evaluating research proposals is a structured and systematic approach that aims to quantify and standardize the assessment process. This method assigns numerical values or points to different components of a proposal, allowing for a more objective and transparent evaluation. Here, we delve into the details of how research proposals are often evaluated using a point system and discuss strategies to enhance proposal scores.

Evaluation of a Research Proposal

14.7.1 Components Typically Evaluated in a Point System

Research Question and Objectives (ten points):

- Clarity and Specificity: Proposals are awarded points for clearly defining research questions and objectives. Precision and specificity in articulating the study's purpose contribute to a higher score.

Literature Review (fifteen points):

- Relevance and Depth: The quality and relevance of the literature review significantly impact the score. Points are awarded for demonstrating a thorough understanding of existing literature and identifying gaps in knowledge.

Methodology (twenty points):

- Rigor and Feasibility: The research methodology is a crucial component. Points are allocated for a well-structured and rigorous design, along with considerations of feasibility, ethical implications, and potential limitations.

Significance and Contribution to Knowledge (fifteen points):

- Innovation and Impact: The proposal's potential to contribute new knowledge or innovative approaches is assessed. Higher scores are assigned for proposals that demonstrate a clear and substantial contribution to the field.

Timeline and Resources (ten points):

- Realistic Planning: The feasibility of the proposed timeline and the adequacy of allocated resources are evaluated. Proposals that present realistic plans receive higher scores in this category.

Research Team and Collaboration (ten points):

- Expertise and Collaboration: Points are awarded based on the expertise of the research team and the potential for collaboration. Demonstrating a well-qualified and collaborative team enhances the proposal's overall score.

Budget and Funding Justification (ten points):

- Clarity and Justification: The budget allocation and justification are scrutinized. Proposals that provide a clear and justified budget allocation receive higher scores in this category.

Communication and Presentation (ten points):

- Clarity and Coherence: The overall clarity and coherence of the proposal's presentation impact the score. Well-structured and articulate proposals are more likely to receive higher points.

14.7.2 Strategies to Improve Proposal Scores in a Point System

Thorough Preparation:

- Understand Evaluation Criteria: Before writing the proposal, thoroughly understand the evaluation criteria and point allocation for each component. Tailor your proposal to address these criteria explicitly.

Clear and Concise Writing:

- Articulate Ideas Clearly: Write clearly and concisely. Avoid ambiguity and ensure that your ideas are presented logically. A well-structured proposal is easier to evaluate and tends to score higher.

Align with Guidelines:

- Adhere to Proposal Guidelines: Follow any specific guidelines provided by the funding agency or institution. Ensure that your proposal addresses all required components and follows the prescribed format.

Emphasize Innovation:

- Highlight Innovation: Clearly articulate the innovative aspects of your research. Whether it's a novel research question, a unique methodology, or an innovative approach, emphasizing innovation can enhance your proposal's score.

Address Feasibility:

- Realistic Planning: Clearly outline a realistic timeline and demonstrate how you plan to manage potential challenges. Addressing feasibility concerns in your proposal contributes to a higher score in the methodology and timeline categories.

Demonstrate Collaboration:

- Highlight Team Collaboration: Emphasize the collaborative nature of your research team. Showcase how the diverse skills and expertise within the team contribute to the overall strength of the proposal.

Justify Budget Appropriately:

- Thorough Budget Justification: Provide a detailed and justified budget. Clearly explain how each budget item contributes to the successful execution of the research. A well-justified budget enhances your proposal's credibility.

Seek Feedback:

- Peer Review: Before submission, seek feedback from colleagues, mentors, or peers. External feedback can help identify areas for improvement and ensure that your proposal is well-received by evaluators.

Adhere to Ethical Standards:

- Ethical Considerations: Address ethical considerations explicitly. Ensure that your research design and methodology adhere to ethical standards, which can positively influence your proposal's score.

Professional Formatting:

- Formatting Matters: Pay attention to the overall professionalism and formatting of your proposal. A well-formatted document reflects attention to detail and contributes to a positive impression.

Remember that while a point system provides a structured evaluation, qualitative aspects also play a crucial role. Striking a balance between quantitative and qualitative strengths in your proposal can lead to a comprehensive and compelling submission. Regularly reviewing and refining your proposal based on feedback and self-assessment will contribute to continuous improvement and increased competitiveness in the evaluation process.

Bibliography

Abdulai, R. T., & Owusu-Ansah, A. (2014a). Essential ingredients of a good research proposal for undergraduate and postgraduate students in the social sciences. *Sage Open*, *4*(3), 2158244014548178.

Abdulai, R. T., & Owusu-Ansah, A. (2014b). Essential ingredients of a good research proposal for undergraduate and postgraduate students in the social sciences. *Sage Open*, *4*(3), 2158244014548178.

Abusaleh, K., & Anwar, A. B. (2022). Meaning and purpose. In *Principles of social research methodology* (pp. 15–28). Singapore: Springer Nature Singapore.

Barker, R. L. (Ed.). (2013). *The social work dictionary* (6th edn). Washington, D.C.: NASW Press.

Best, J. W., & Kahn, J. V. (1986). *Research in education* (5th edn). New Delhi: Prentice Hall.

Bonnel, W., Smith, K., & Gifford, J. M. (2015). *Proposal writing for nursing capstones and clinical projects*. New York: Springer Publishing Company.

Boyack, K. W., Smith, C., & Klavans, R. (2018). Toward predicting research proposal success. *Scientometrics*, *114*, 449–461.

Bryman, A. (2016). *Social research methods* (5th edn). Oxford, United Kingdom: Oxford University Press.

Cohen, L., Manion, L., & Morrison, K. (2018). *Research methods in education* (8th edn). London: Routledge.

Connaway, L. S., & Powell, R. R. (2010). *Basic research methods for librarians*. London: ABC-CLIO.

Creswell, J. W. (2008). *Educational research: Planning, conducting and evaluating qualitative and quantitative research* (4th edn). Upper Saddle River, NJ: Pearson Education, Inc. Hall.

Creswell, J. W., & Creswell, J. D. (2017). *Research design: Qualitative, quantitative, and mixed methods approaches* (5th edn). London & New York: Sage.

Creswell, J. W., & Poth, C. N. (2016). *Qualitative inquiry and research design: Choosing among five approaches*. London & New Delhi: Sage.

Day, A. R. (2018). *How to write and publish a scientific paper* (8th edn). Cambridge University Press.

Denscombe, M. (2012). *Research proposals: A practical guide*. UK: McGraw-Hill Education.

Gastel, B., & Day, R. A. (2022). *How to write and publish a scientific paper*. California: Bloomsbury Publishing USA.

Gerin, W., Kapelewski Kinkade, C., & Page, N. L. (2012). *Writing the NIH grant proposal: A step-by-step guide.* London: Sage.

Given, L. M. (Ed.). (2008). *The Sage encyclopaedia of qualitative research methods.* London & New Delhi: Sage.

Hansen, E. C. (2020). *Successful qualitative health research: A practical introduction.* Routledge.

Henn, M., Weinstein, M., & Foard, N. (2009). *A critical introduction to social research* (2nd edn). London, California & New Delhi: Sage.

Indeed Editorial Team. (2023). *What is a research proposal? (Plus how to write one).* <https://www.indeed.com/career-advice/career-development/research-proposal>

Islam, M. R. (2019). Designing a PhD proposal in qualitative research. In M. R. Islam (Ed.), *Social research methodology and new techniques in analysis, interpretation, and writing* (pp. 1–22). Pennsylvania, USA: IGI Global.

Islam, M. R. (2022a). Inquiry: A fundamental concept for scientific investigation. In M. R. Islam, N. A. Khan & R. Baikady (Eds), *Principles of social research methodology* (pp. 3–14). Singapore: Springer Nature Singapore.

Islam, M. R. (2022b). Preparation and development of data collection instruments for social research. In M. R. Islam, N. A. Khan & R. Baikady (Eds), *Principles of social research methodology* (pp. 449–461). Singapore: Springer Nature Singapore.

Islam, M. R., & Banda, D. (2011). Cross-cultural social research with indigenous knowledge (IK): Some dilemmas and lessons. *Journal of Social Research & Policy,* 2(1), 67–82.

Islam, M. R., & Hajar, A. B. S. (2013). Methodological challenges on community safe motherhood: A case study on community level health monitoring and advocacy programme Bangladesh. *Revista de Cercetare si Interventie Sociala,* 42, 101.

Islam, M. R., Cojocaru, S., Hajar, A. B. A. S., Abd Wahab, H., & Sulaiman, S. (2014). Commune and procedural level challenges and limitations in conducting social research in Malaysia: A case of disabled people. *Revista de Cercetare si Interventie Sociala,* 46, 255.

Islam, M. R., Khan, N. A., Ah, S. H. A. B., Wahab, H. A., & Hamidi, M. B. (2021a). Introduction to the field guide for research in community settings: Tools, methods, challenges and strategies. In M. R. Islam, N. A. Khan, S. H. Ah, H. A. Wahab & M. B. Hamidi *Field guide for research in community settings* (pp. 1–10). Cheltenham, UK: Edward Elgar Publishing.

Islam, M. R., Khan, N. A., Ah, S. H., Wahab, H. A., & Hamidi, M. B. (Eds). (2021b). In M. R. Islam, N. A. Khan, S. H. Ah, H. A. Wahab & M. B. Hamidi *Field guide for research in community settings: Tools, methods, challenges and strategies.* Edward Elgar Publishing.

Jansen, D. (2020). *What (exactly) is a research proposal?* https://gradcoach.com/what-is-a-research-proposal-dissertation-thesis/

Karim, M. R. (2022). Designing research proposal in quantitative approach. In M. R. Islam, N. A. Khan & R. Baikady (Eds), *Principles of social research methodology* (pp. 131–156). Singapore: Springer.

Khan, K. K., & Mohsin Reza, M. (2022). Social research: Definitions, types, nature, and characteristics. In M. R. Islam, N. A. Khan & R. Baikady (Eds), *Principles of social research methodology* (pp. 29–41). Singapore: Springer Nature Singapore.

Kothari, C. R. (2004). *Research methodology: Methods and techniques.* New Delhi: New Age International (P) Limited Publishers.

Kumar, A. (2002). *Research methodology in social science.* New Delhi: Sarup & Sons.

Kumar, R. (2011). *Research methodology: A step by step guide for beginners* (3rd edn). New Delhi: Sage.

Leedy, P. D., & Ormrod, J. E. (2015). *Practical research: Planning and design* (11th edn). Essex, England: Global Edition.

Leedy, P. D., & Ormrod, J. E. (2019). *Practical research: Planning and design* (12th edn). London: Pearson.

Locke, L. F., Silverman, S. J., & Spirduso, W. W. (2019). *Proposals that work: A guide for planning dissertations and grant proposals* (7th edn). London: Sage.

Marczyk, G. R., DeMatteo, D., & Festinger, D. (2010). *Essentials of research design and methodology* (Vol. 2). Hoboken, NJ: John Wiley & Sons.

May, T. (2011). *Social research: Issues, methods and process* (4th edn). Maidenhead, Berks: Open University Press.

Mishra, D. S. (2017). *Handbook of research methodology: A compendium for scholars & researchers.* New Delhi: Educreation Publishing.

Myers, J. L., Well, A. D., & Lorch Jr, R. F. (2013). *Research design and statistical analysis.* New York: Routledge.

Neuman, W. L., & Robson, K. (2018). *Basics of social research: Qualitative and quantitative approaches* (4th edn). Canada: Pearson Canada Inc.

Penz, E. (2006). Researching the socio-cultural context: Putting social representations theory into action. *International Marketing Review, 23*(4), 418–437.

Punch, K. F. (2016). *Developing effective research proposals.* London: Sage.

Reddy, C. D. (2019, June). Thinking through a research proposal: A question approach. In *18th European conference on research methodology for business and management studies* (p. 271). Johannesburg.

Russell, S. W., & Morrison, D. C. (2007). *The grant application writer's workbook: NIH version.* Massachusetts, USA: Jones & Bartlett Learning.

Sarantakos, S. (2017). *Social research.* London: Bloomsbury Publishing.

Smith, G. W., Mykhalovskiy, E., & Weatherbee, D. (2006). A research proposal. In Smith, D. E. (Ed.), *Institutional ethnography as practice* (pp. 165–180). New York: Rowman & Littlefield Publisher.

Sternberg, R. J. (2012). *Writing successful grant proposals from the top down and bottom up.* London & New Delhi: Sage.

Tabatabaei, F., & Tayebi, L. (2022). *Research methods in dentistry.* Switzerland: Springer Cham.

University of Birmingham. (2023). *How to write a research proposal.* https://www.birmingham.ac.uk/schools/law/courses/research/research-proposal.aspx.aspx

Vasanthakumari, S. (2021). Writing research proposal. *World Journal of Advanced Research and Reviews, 10*(1), 184–190.

Wolery, M., & Lane, K. L. (2014). Writing tasks: Literature reviews, research proposals, and final reports. In Ledford, J. R., & Gast, D. L. (Eds), *Single case research methodology* (pp. 50–84). New York: Routledge.

Yin, R. K. (2009). *Case study research: Design and methods.* New York: Sage.

About the Author

M. REZAUL ISLAM currently holds the position of Professor of Social Work at the Institute of Social Welfare and Research, University of Dhaka, Bangladesh. Previously, he served as a Visiting Senior Lecturer in the Department of Social Administration & Justice at the University of Malaya, Malaysia, from December 2012 to December 2016. Dr Islam earned his Master of Social Work (MSW) and PhD degrees from the University of Nottingham, England. With a career spanning over twenty-seven years, Dr Rezaul has been actively engaged in advancing research opportunities through international collaborations with various universities and institutions globally. He has delivered numerous lectures, public talks, and research presentations at universities worldwide. His expertise encompasses a wide range of subjects, including health and well-being, community development, family and child care, international and comparative social policy, climate justice, international labor migration, poverty and social inequality, and social change and globalization. Dr Rezaul has an extensive publication record, with 130 journal articles, 50 book chapters, and 15 books published by reputable international publishers. He currently serves as a member of the International Advisory Board of the Community Development Journal (Oxford University Press) and holds editorial board positions for four journals: Asian Social Work and Policy Review (Wiley), Local Development & Society (Taylor & Francis), International Journal of Community Well-being (Springer), and SN Social Sciences (Springer). In addition to his academic role, Dr Rezaul has collaborated extensively with international organizations such as the World Bank, UNDP, UNICEF, ILO, ADB, British Council, Plan International, Concern Universal UK, Concern Worldwide, and others. He has successfully completed approximately thirty international research projects, where he served as either a team leader or chief investigator.

Index

academic integrity xii–xvi, 8, 187, 192
accuracy 26–27, 128, 130, 132, 134, 184
achievable xv, 14, 52, 54, 59, 60, 78, 80, 83, 87, 100, 159, 161, 164, 171–172, 194
adaptability 60, 169, 204
anticipated results 159, 163

background 6, 14, 16, 25, 33, 37, 39–41, 43–45, 47, 49, 50, 53, 55, 64, 84, 94, 134
biases 41, 56, 61, 64, 99, 112, 133, 148–149, 151, 200, 202
boundaries xv, 11, 14, 42, 106, 113, 131, 153–157, 194, 203
budget xvi, 9, 14, 16, 132, 155, 156, 169, 174–175, 178–181, 184–185, 193, 210–211

case study 111, 118, 159
communication ix, 6, 10, 25, 27, 62, 119–120, 138, 142, 147, 162, 169, 183, 194, 196–197, 202, 204, 207, 210
conceptual framework 69–72, 74–75, 90, 94, 109, 206
confidentiality xv, 61, 63–64, 122, 127, 135–137, 139–140, 143, 145, 150–152, 155–156, 201
culture 118, 146

data analysis 8, 13–14, 16, 43, 48, 53, 61, 71, 83, 96, 101, 111, 114–115, 128–132, 134, 146, 149–150, 162, 170–171, 174, 177–178, 182

data collection xv, 8, 13–14, 16, 48, 53, 59, 61, 70, 77–78, 83, 86, 91–92, 96–97, 99, 101, 109, 111–112, 114, 116, 120–128, 131–134, 136, 139, 145–146, 147, 149, 154–156, 162, 170–175, 180, 194, 206
data collection instrument 111
data interpretation 83
data management 145–146

expected outcomes xvi, 9, 14, 54, 71, 159–165, 167

feasibility 5–6, 17, 47, 59, 67, 81, 87, 89, 97, 100–102, 108–109, 127, 155–156, 161, 163–164, 169, 178, 184, 193–195, 205–207, 209, 211

Gantt chart xvi, 169–174
gap identification 40, 42, 50

hypotheses 2, 11, 42, 70–71, 73, 75, 78, 84, 89, 91, 94–97, 114, 160, 163

impactful construction 78
importance of research proposal 1, 10
informed consent xv, 8, 61, 63, 122, 127, 135–137, 139, 142–144, 150–152, 155–156
integrity xii–xiii, xiv–xvi, 12, 31, 57, 61–62, 64, 137, 139, 145–148, 187, 192, 194, 201, 203
introduction xiv, 1, 6, 14, 16, 30, 32, 35, 37–45, 75, 79, 84

jargon words 19
justification 5, 11, 37, 41–42, 54, 64–66, 89, 91, 94–95, 99, 112, 144, 161, 180, 184, 210–211

key components xvi, 14, 154, 170, 193, 195, 198
keywords 1, 1, 19–21, 23–24, 37, 47, 69, 77, 89, 111, 135, 153, 159, 169, 187, 193

literature review xi, xiv– xv, 3, 6, 8, 14, 16, 18, 39, 41, 43, 48, 49–50, 65, 69–71, 74, 83, 89–90, 97–100, 109, 125, 170–171, 173–175, 187–188, 206, 209

methodological overview 37, 43
methodology xi, xiv–xv, 5, 8–9, 14–18, 22, 28, 34, 48, 53–54, 59, 69–70, 74, 97, 99, 101, 111–114, 116, 126, 131–132, 149, 151, 154, 156, 162, 164, 173, 175, 184, 194, 206, 209, 211–212

online education 159, 165–167
opening statement 37–39, 43

parameters 153–154
participant welfare 135, 144
peer review xvi–xvii, 58, 148, 150, 193, 197–203, 212
    peer reviewers xiii, xvii, 10, 15, 148, 199
policy link 89, 106, 108
problem statement xiv, 9, 48–52, 54–67

qualities 169, 184–185
qualitative dimensions 193
quantity 4, 183, 187–188
quantitative assessment 193

rationale of the study xv, 83, 89–91, 98, 108
realistic 13, 58–60, 62, 78, 81, 83, 97, 100, 159, 161, 172, 185, 194, 209, 211
reference styles 187, 190
relevance xiii, xv, 6, 19–20, 24, 32, 34–35, 37, 40–41, 47, 49, 50–51, 53–56, 65, 67, 79, 82, 87, 89, 91–95, 99, 102, 104, 106–107, 112, 127, 155, 157, 161–162, 164, 187, 193, 195, 197–198, 209
reliability 61, 70, 73, 103, 111, 119, 122–123, 125–126, 128, 130, 134, 136–137, 145, 194, 199
research approach xv, 3–4, 16, 111, 114–117, 119–120, 122, 133, 136
research context 31, 94, 101
research contribution 19
research design 1, 8, 14–18, 43, 53, 59–61, 71, 74–75, 77, 81, 83, 87, 97, 101, 109, 131–132, 144, 152, 154, 156, 162, 164, 194, 206, 212
research ethics xv, 47, 58, 62, 64, 135–138, 152
    ethical review process 180
    evaluation criteria xvi, 193, 195, 197–198, 202, 210
research impact 19
research methods 17, 83, 92, 109, 111, 116–119, 124, 134
research objectives xi, xiv–xv, 1, 8, 10, 16, 31, 37, 42–44, 47–48, 52–54, 59, 75, 77–83, 86–87, 89, 95–97, 99–100, 109, 112, 114, 116–117, 119, 122–128, 154–157, 160, 162–163, 167, 188, 193
    general objective 78–80
    specific objective 79, 81
research problem xiv, 15, 32, 37, 39–44, 47–58, 64, 70, 74–75, 78–82, 85–87, 109, 112, 116, 194

Index

research questions xv, 6–8, 11, 13–16, 22, 32, 42, 54, 62, 69–71, 73, 75, 77–78, 82–87, 90–91, 94–95, 97, 101, 109, 114, 116–117, 119–120, 122, 124–125, 127–129, 131, 154, 160, 162–163, 206, 209
research title xiv, 19–35
resources ix, xvi–xvii, 6–7, 12–13, 48, 51, 53, 58–61, 81, 83, 87, 100–101, 103, 112, 116, 127, 148, 156, 161–162, 164, 169, 171, 172, 174, 178–179, 194, 209

sampling 16, 99, 101, 111, 115, 125–128, 134, 146
scholarly discourse 187, 198
scope of the research xv, 9, 42, 77, 131, 153–154, 157, 187

sensationalism 19, 27–28
societal impact 89, 104, 106, 109, 138
strategic vision 159
student learning outcomes 160, 165, 167
sufficiency 187

theoretical framework xiv–xv, 8, 14, 19, 22, 29, 32–33, 69, 72, 74, 78, 86–87, 92, 98, 103, 130
transparency 28, 58, 61–62, 64, 138, 141, 146, 148–149, 153, 159, 162, 169, 178–179, 184, 197–198, 200, 202–203
types of research proposal 1, 7

validity 57, 61–62, 70–71, 73, 111, 113, 116, 119, 122–123, 125, 126–128, 130–132, 134, 145, 194, 197